Other books by the author:

*Grand Canyon: A History of a Natural Wonder
and National Park.*

THE SHADOWS MOVING IN THE MOON'S SKULL EYES:

A VISION OF APOLLO XI

DON LAGO

LIVINGSTON PRESS
THE UNIVERSITY OF WEST ALABAMA

Hardcover binding by:
Typesetting and page layout: Sarah Coffey
Proofreading: Sarah Coffey, Nick Noland, Joe Taylor

Cover layout: Amanda Nolin and Sarah Coffey
Cover art:
AS11-40-5961
Armstrong photo of LM from a distance All photographs on
this website are courtesy of the
National Aeronautics and Space Administration, specifically
the
NASA History Office and the NASA JSC Media Services
Center.
Back cover art: jsc2018e003251 (Jan. 31, 2018)–The lunar
eclipse "Blood Moon" was photographed from
the Johnson Space Center in Houston,Texas, during the early
morning hours of Jan. 31, 2018.
Credit: NASA/Robert Markowitz

Livingston Press is part of The University of West Alabama,
and thereby has non-profit status.
Donations are tax-deductible.

first edition
6 5 4 3 3 2 1

THE SHADOWS MOVING IN THE MOON'S SKULL EYES:

A VISION OF APOLLO XI

CONTENTS

PROLOGUE:
THE ECLIPSE

The craters stared at me, the empty eyes of thousands of skulls. The craters were full of shadows and fierce violence. The craters stared blindly, stared into the gulf between death and life, stared into my eyes still plump and blue and pulsing with light and shapes, yet stared without recognition.

I looked into the dimness, into the huge chamber, up and down the chaotic pile of tens of thousands of bones and skulls. Many skulls held cracks or jagged holes or were shattered into half skulls. Many skulls were missing jawbones, and many jawbones were missing some or all of their teeth. Thousands of skull fragments lay scattered randomly. Many of the bones were cracked or pitted or snapped apart, leaving splinters. Rib bones and pelvis bones announced curves but abruptly canceled them. Thousands of finger and toe bones had settled to the bottom of the pile. In places, the bones had been sorted and stacked, a section of arm bones, a section of leg bones, but then the bones seemed to recognize the pretense of such order and the honesty of chaos and became a jumble of every kind of bone, pointing every direction and no direction. The skulls seemed to have ended up wherever they would fit, a few here, a few dozen there, a line of skulls here, chaos there, skulls right-side up or upside down, pointing in every direction and no direction. The craters that had once held eyes and sparkled with moonlight now conducted only dimness and bone dust.

These were the skulls and bones of 130,000 men. At least, this was the official guess. It was hard to guess when the bodies had been hit by millions of explosions and shattered and tossed apart: skulls flew away from bodies, and bones from different bodies got mixed together only to be hit again and shattered further and separated again. Some sixty million artillery shells had landed here, digging sixty million craters large and small, deep and shallow, circular and oblong, spraying out dirt that buried men alive and filled up craters dug only minutes or hours before. A new explosion exposed dozens of bones, exposed them to the sun for a day, exposed them to the night to answer the whiteness of the moon. The explosions continued for nearly a year. The bombardment shattered bedrock that had endured for tens of millions of years and turned it into new soil. Twenty-year-old human bodies had no chance to resist; their flesh and blood and excrement made much easier soil. The 130,000 were so thoroughly shattered that no one could identify them, not even which army they had belonged to. So they were entombed together, comrades and enemies embracing, with the identities that had made them hate one another entirely erased.

Some two and a half million men had fought here, fought back and forth, back and forth, back and forth for nearly a year, and some third of a million men had died here. This number too was a guess. A hundred thousand men had simply disappeared. Had they sneaked away in the night and gone home, or had they been so thoroughly shattered and buried that no one had ever accounted for them? Every year for nearly a century, spring rains eroded from these hillsides and valleys new bones, so perhaps another hundred thousand skeletons still lay unfound. Many of the 130,000 skeletons had been unearthed by accident, by foresters digging holes to turn the barren battlefield into a forest; no one had ever systematically excavated the entire battlefield for bones.

I was in the Douaumont Ossuary atop the highest ridge at the battlefield of Verdun in France, not far from the German border. Never in the annals of human warfare had so many given so much for so little. Verdun had defined World War One and its western front and trench warfare, where national pride and antique military tactics collided with the brutal new realities of industrialized warfare. Upon a ridgeline with only modest strategic advantages, both sides affixed their pride, their identities as superior nations, and thus neither side could bear to give ground and both sides poured in lives and arms. When the war began, these skulls had beamed with delight and certainty for the glory they would win, and these arm bones had waved hats and flags and rifles. At least they got this art-deco ossuary, with two long corridors, a chapel, an eternal flame, a grand tower (with a round top that suggests an artillery shell), and more than a dozen chambers packed with anonymous skulls. Unlike church ossuaries and catacombs, where the bones were arranged carefully and even turned into works of art, Verdun's ossuary was dedicated to chaos.

I had come here to witness a total eclipse of the sun, in 1999.

Verdun was not my first choice from where to watch the eclipse. I had studied the eclipse map, with its narrow band of totality running across Europe, and I noticed that it included Ulm, Germany, the birthplace of Albert Einstein. Einstein seemed the perfect patron saint for an eclipse. Out of his cradle, out of his brain had unfolded the geometries of the universe, of space and time and forces strange and powerful, including the geometries the planets and moons trace out. In a 1919 total solar eclipse, British astronomer Arthur Eddington observed Mercury crossing the sun and confirmed that Einstein's general theory of relativity offered a more accurate map of gravity and planetary motions than Isaac Newton's theories, lifting Einstein out of the cloister of physics and making him an international celebrity. Part of the news sensation was that not long after British and German troops had been killing one another, a British scientist was elevating a German scientist above his own national icon, Isaac Newton. In a world exhausted from nationalism, astronomy had risen higher.

At Ulm, totality would last two minutes and five seconds. I imagined myself standing on some cobblestone street outside the house where Einstein was born, where he had first been wonderstruck by the night sky or the mysteries of magnets. But as I investigated further, I learned that Einstein's house was no longer there. His street had been bombed during World War Two. The bombardier probably didn't notice that his bombs were falling according to Einstein's gravity and not Newton's. I decided that Einstein's ghost house would still work, but when I got to Ulm I found that his street was narrow and sky-obstructed, so I decided to watch the eclipse from the nearby, huge plaza in front of the Ulm cathedral, with the world's tallest cathedral spire pointing skyward. Ulm seemed even more appropriate when I discovered that its fourteenth-century city hall held an elaborate astronomical clock, including lunar cycles.

I had several days before the eclipse, so I set off on a cultural tour of this region of Germany. I visited Weimar, where Goethe contemplated humankind's Faustian spirit and fate, where Schiller offered his Ode to Joy, and where Nietzsche considered whether humans without God would become supermen or madmen. I drove up the ridge, not many miles, within smoke-sniffing distance, to where the Nazis had built the Buchenwald concentration camp for Germany's political, scientific, religious, and cultural elite (including Elie Wiesel). It was to Buchenwald that Wernher von Braun went to find prisoners better qualified to build rockets than the common laborers at his Peenemünde prison camp; the SS transferred 636 prisoners to von Braun. It was on the hillsides around Buchenwald, the same hillsides where Goethe had sought nature's inspiration, that the boy von Braun had attended school and become possessed by the idea of traveling into space. At the end of the war, American generals, shocked by what they were discovering inside the concentration camps, ordered that 1,200 residents of Weimar be marched up the ridge to Buchenwald to see the piles of corpses, the ovens, and Nazi "souvenirs" like shrunken heads and skin tattooed with German knights.

I visited Bayreuth, where the annual Wagner festival was underway, loudly glorifying German gods and heroes and warrior valor, drawing tens of thousands of pilgrims from around the world, even Jews carrying the still-bold idea that the universe is ruled by benevolence.

The day before the eclipse, the time for me to head back to Ulm, I began studying the weather forecasts. For days the sky had been overcast, sometimes rainy. The forecasts predicted more clouds and more rain for eclipse day. For hours I watched the morning television newscasts, switching from channel to channel, from country to country, from language to language, none of which I could understand, studying the maps to get the best information and to make the best decision about where to go. I would need to decide this morning and make a run for it, but which way? If I headed east into Austria I'd be getting into mountains and probably worse weather, and the mountain roads would offer less flexibility for making a last-hour chase for a break in the clouds. France seemed to offer a lower chance for rain, if still too much. I got out my eclipse map and studied the path of totality in eastern France, and I saw that it included Verdun. I didn't know much about the geography of France, but I knew that generals everywhere liked to take and hold the high ground. A bit of high ground might be helpful, not only for viewing the sky but for seeing the lunar shadow moving across the land. I thought of American Civil War battlefields, public places with miles of roads and open fields; if Verdun was similar, I wouldn't need to trespass on some farmer's field or driveway. I headed for Verdun.

I did not know that Verdun was one of the rainiest places in France, with rain about half its days. When I woke up the next morning, I didn't need to peek outside my tent to tell there was no sunshine. The eclipse was going to start in three hours, at 9:10 a.m. I drove to the battlefield and found that plenty of other people had decided it was a good place for seeing the eclipse. In front of the ossuary was a large plaza facing the sun, the would-be sun, and a crowd was gathering, some with telescopes and sophisticated cameras. On the lawn sloping below us, 16,142 crosses marked the graves of soldiers who had been identified, but only French soldiers.

The rain started. I took shelter in the ossuary. I stared into eyes whose sun had been eclipsed long ago.

The battle was eighty-three years ago, so a few of these skulls could have been alive today, sitting on porches and telling tales. These skulls had been capsules heading toward the future on paths not as reliable as 9:10 a.m. planetary orbits but still capable of carrying them for a century. These skulls would have glowed with women's faces and children's joy and Christmas lights. But all these capsules had crashed here, stopped tree-ringing here. All these bones rising toward the future became the broken and fallen columns of a ruined city. The skulls told me how they had cared about nation and domination more than they valued life itself.

The rain let up, but the clouds did not. I went for a walk through the battlefield. It remained miles of craters, a million craters. The farmers who had owned this land made little effort to reclaim it after the war, for it remained too rough to plow or plant and held thousands of still-unexploded artillery shells and too many corpses. Who would want to buy food juicy with the blood and gore of dead soldiers? Some hilltops held the ruins of cement forts and the traces of trenches and tunnels. When I wandered off the pathways and into the fields and woods, my boots quickly clotted with mud, mud no doubt rich with lost blood and pulverized flesh.

Interpretative panels held many photos of the battlefield soon after the battle, cratered landscapes that resembled the moon. More than one soldier was quoted comparing the battlefield with the moon.

Now the moon would descend here again. A landscape of craters would come; a landscape without life would come; the shadow of death would invade Earth again.

At the time the eclipse was supposed to begin, when the edge of the moon first touched the sun, the sky was still thoroughly cloudy. It seemed we were going to miss the eclipse. We waited, ten minutes, twenty minutes, thirty minutes, but the clouds remained. We waited. The whole eclipse, from first contact until last, would last nearly three hours. We waited as only Earth could wait, not as a moon that had waited four and a half billion years without even knowing it was waiting, counting time only with the hourglasses of boulders decaying into dust. We waited as only life could know the passage of time and the coming of events, metabolizing time into feelings, into hope and plans and worry for the future, into recognition of the motions of forces far larger than life, into awe at the workings of the universe.

The sky and the ground were unusually dark: was this because the clouds were hopelessly thick, or because the advancing eclipse was blocking the light?

Forty-five minutes before totality, the clouds began to thin and break. We caught glimpses of a sun with a large slice covered. (Glimpses, of course, through protective glasses, for even in an eclipse the sun remained far more powerful than us, more powerful than eyes that without thanks used the sunlight to mark their territories and aim their artillery shells). Then the clouds parted, like curtains opening. We saw a huge arc of clear sky. The crowd cheered. As totality approached, clouds occasionally crossed the sun, but totality itself would be entirely clear. Only after the eclipse ended did the clouds move back in, covering the sky for the rest of the day. We would be among the few in Europe to see the eclipse. The crowd cheered, and not just for the gift of good weather.

The ground and the air held an unusual dimness and glow, not from the weather but entirely from the eclipse. The moon continued advancing, too slowly for human eyes to see it moving from moment to moment, but minute-hand fast enough to notice it advancing from minute to minute. The moon, the expert on phases, showed the sun how to be a crescent, four-fifths whole, two-thirds whole, a half sun, less than half. Slowly, steadily, the clockwork solar system revealed how it kept giant balls rolling in perfect motions, perfect orbits; silently, so silently, the music of the spheres played its massive harmonies. For once, I could feel the ground beneath me rolling through space, synchronized with the moon and sun. Occasionally I took my eyes from the sky and looked at the ground and saw it growing dimmer, its color changing, more silver, or sort of grey, or maybe blue. I noticed the grass and trees glowing oddly, and I held out my hand and saw it too glowing oddly, which of course only revealed the strangeness we had possessed all along but refused to notice. I noticed my shadow fading away, merging into the greater shadow, the cosmic night, to which we had belonged all along.

As the light faded, all the plants around me, the grass and trees and cemetery flowers, began going hungry—an ancient, deeply rooted hunger. The energy flowing within them slowed, and their biological looms began shutting down their merging of water and air and soil into molecules and cells and plant forms. Tricked into thinking night was arriving, they began relaxing their leaves and fading into sleep.

Birds stopped singing and some darted about in apparent confusion, perhaps heading home to their nests; these birds were the descendants of birds that had once fled the thunder of cannons. Insects too went quiet. I looked at a couple of pet dogs in the crowd and they too appeared a bit nervous, perhaps from the same instinctive animal sense of normality and violation that set humans imagining dragons devouring the sun and set them banging drums and performing magic to defend the sun.

As the moon advanced, the air grew noticeably cooler. The coldness of space was reaching through and touching our faces.

The intercepted sunlight, the light that would have been nourishing the life of Earth, was now falling onto a world of nothing but rock and dust, craters and lava and rubble. The light was not meeting any air that mellowed it into a pleasant day and unfolded its secrets into a colorful sky, into rainbows or sunsets; now the light meant intense heat. The light did not sparkle with the energy of oceans and rivers and lakes nor inspire them to rise into clouds and rain. The only greenness the light found was in specks of volcanic glass. The light found mostly greyness. The light created only crude shadows, for it found only crude shapes. The light set the molecules of the dust buzzing with energy but the dust lacked the forms and skills to do anything with this energy and only threw it away. The dust didn't know the secrets of using the light to turn formlessness into form, to weave land and water and air into coral reefs and forests and dolphins. Energy that would have become the galloping of horses and the flapping of wings had nowhere to go. Light that would have set eyes aglow was not seen by anyone. Light that would have become the light of consciousness, that would have glowed with wonder at the cycling of Earth and moon and sun, ended up as more littered energy in a universe of chaos and obliviousness.

The whole universe is like the moon, chaotic and lifeless and mindless. This is a universe of craters, of planets and moons saturated with craters, craters within craters, craters atop craters, craters ruining other craters. This is a universe of rubble thrown out of craters, of boulders and fragments and vast plains of dust. This is a universe of rocks attacking one another, of asteroids and comets colliding with one another and with planets and moons. This is a universe of gas, of huge clouds of gas and planets of gas. This is a universe of ice. This is a universe of too much cold and too much heat, too much darkness and too much radiation, too much overcrowded matter and far too much emptiness. This is not a universe of life, not much of it, only isolated specks of it, not a universe where life should regard itself as normal or inevitable or safe. The universe is enormously talented at blocking and extinguishing life. Throughout the universe, intelligent species have watched helplessly as a rogue star wandered toward them, turning tides into tsunamis, cracking tectonic plates and setting volcanoes flowing, ruining ancient cities. Civilizations have watched nearby stars going supernova, watched nearby black holes growing larger and more energetic. Worlds full of life have been engulfed and burned to dust, and their planets have entirely disappeared. Worlds long full of life have been severely disrupted by asteroid impacts, by mobs of volcanoes filling the atmosphere with poisons, by climates deteriorating.

So why is it that in an ungenerous universe of craters, where Earth had been set free from the curse of craters, why is it that I am surrounded by a million craters? Why has the life of Earth chosen to inflict upon itself the chaos of the universe? To paste upon Earth the deadness of the moon? Why have humans devoted so much talent and energy to cramming supernovas into little packages that we can drop upon ourselves?

It was because of the way we defined "ourselves," or failed to define it. We didn't perceive ourselves as simply "life," the same rare force wearing many different forms and faces. We defined ourselves by our separateness, by the varied packages into which we had been divided. We defined ourselves by where we had been born, by differences in appearance or tribal stories or how we shaped sounds, by our favorite foods or music, by our passion for being different. We had inflicted upon ourselves the craters in our own identities. The chaos of the universe, of crashing worlds and exploding stars, had manifested itself in a new way, as a dysfunctional orbit of imagination. When you combine our failure to identify ourselves as life, our antibody-like impulse of hostility toward alien forms, our preference as social animals for identifying ourselves with the roles of societies and hierarchies, and our valuing status, pride, power, and domination more than we care about being alive, it leaves humans far too ready to destroy life, including our own lives.

The moon continued moving across the sun; the sky and ground darkened more noticeably; the air continued cooling. Strange pulses of light flickered across the ground and upon the ossuary tower as the last rays of sunlight were filtered by Earth's atmosphere. When the sun was nearly covered, nearly a black disc, its edges held a few flickers of light, beams of light flowing through deep valleys on the moon, the unphilosophical moon that had never been the perfect circle humans had once wanted it to be. The last of these beams sparkled for a moment, reminding humans of their own realm of experience, which they imposed upon the platonic sky by calling it "the diamond ring effect." Then the diamond went out and left only the ring itself, the solar corona, a thin aura of light around the darkened sun, an aura of red prominences whipping outward. The universe might not consist of perfect spheres moving in perfect circles, but this eclipse did reveal the uncanny perfection of a moon that had just the right size and distance to cover the sun completely; millions of years from now, as the moon continued inching away from Earth, the moon would no longer perform total eclipses.

In the darkened sky, a few planets came out.

I stood engulfed by the moon's shadow, embedded in the cycles of the solar system, carried along in a vast symphony of motion. Earth was revealing that it had a larger identity than being a mere stage for puny human stories. I stood in a cosmos inviting me to see my own larger identity. I stood in awe. I felt an acute sense of time, of two minutes passing more intensely than almost any minutes I had ever experienced, knowing that the eclipse would soon be over and that I might never see another one. I urged myself to rally attention, to absorb this event before it was gone.

When the sun went dark, the crowd gasped and aaa-hed and cheered. I would guess that in all their decades this was the only time the ossuary and cemetery had heard a cheer. This was not the insane cheering with which these skeletons had welcomed the coming of war. This was a cheer for life. People were cheering not simply for their own luck at seeing the eclipse, but because the far less reliable astronomical weather and biological weather had parted and allowed Earth to become alive and become us. This time the lunar landscape had descended here not from wrath but from the cycles of nature, not from human smallness but from cosmic largeness, and it was offering us a contrast between moon and Earth, between chaos and order, between death and life, a contrast that showed us how small is the house of life in the universe. People were cheering as if in rebuttal to the silence of the graves around us, as if lunar craters could not remain a final verdict on the value of human life. We cheered for the fertile Earth, for the sun that gave it life, and for a universe more faithful to us than we were to ourselves.

Onward the moon advanced, showing its diamond ring again, then breaking its marriage with the sun. Again the sunlight flowed through lunar valleys, this time on the moon's opposite rim. Onward the moon shadow advanced across the ground, flowing away from us, letting the ground brighten again. Out of the darkness of the cosmos emerged human shadows, shapes vague at first yet soon evolving heads and hands, a baby being born, shadows that preserved some of the cycles and secrets of the moon and cosmos. And yet as I watched my shadow growing, I again became a separation, an entity apart from the whole cosmos and all other separate forms. Unlike the moon shadow, this shadow was entirely under my control; I could make its hand wave or its head move.

As I watched the moon drift across the sky, making the sun a crescent again, I thought of the footprints on the moon, the footprints of a dozen separate humans. I saw half a dozen flags proclaiming the separateness of nations and the superiority of one. I saw a plaque offering what for humans was an unusually generous and noble sentiment: "We came in peace for all mankind." Yet this wasn't entirely true, and everyone knew it. Humans had raced to the moon mainly out of national pride, out of the same drive for superiority that had turned Verdun into a lunar landscape.

And yet, something a bit surprising had happened along the way. Even soldier-astronauts steeped in nation and domination, even thrill-seeking pilots who wanted to fly faster and higher than anyone else, had looked upon the cratered moon and looked back upon the tiny blue Earth and been struck by the contrast, by the difference between a dead world and a living world. Even astronauts who disavowed any poetic tendencies didn't do a bad job of pointing out this contrast and its importance.

Through the moon's grey, cratered mirror, we might finally be able to see ourselves clearly.

1
OUT OF CHAOS

Out of the darkness, out of the emptiness, out of the brilliant yet oblivious sky, out of the chaos from which asteroids arrived randomly and inflicted their chaos, toward a ground of craters and rubble and dust, a star appeared among the unwavering stars and sparkled and moved, descending slowly, coming to show the chaos what had been happening elsewhere over the last four billion years.

For billions of years asteroids had crashed onto the moon and blown out wounds and smashed themselves apart. The moon had never noticed or cared. The asteroids had never noticed or cared. The moon and the asteroids had always been oblivious of their own existence and their own identities, always been loyal to a cosmos of oblivious bodies following the commands of forces more powerful than themselves, forces inside themselves and outside themselves, a cosmos of stars and nebulae and planets and moons forming and enduring and disbanding without ever knowing what or why. The moon and the asteroids took no evasive actions to avoid colliding; with their gravity and momentum, they drew themselves together. Even when an asteroid smashed itself into dust it did not entirely lose its identity, for its primary name had always been Chaos and now it remained Chaos in another form. Even when the moon received a large and permanent scar it also gained more mass and more identity as Chaos.

The asteroids came in a chaos of shapes, sizes, compositions, solidities, colors, speeds, spins, and trajectories. A few asteroids were nearly round, but never precisely round. Most were quite irregular: oblong, lumpy, jagged, ridged, cracked, broken, thoroughly cratered. Some were miles across, flying mountains, but most were a foot or two wide. Some were made of metal, some of stone, and some were mixtures of metal and stone. Some were solid bodies but more were loosely bound. Some were as dark as space, while others stood out. Some came in at ferocious speeds, tens of thousands of miles per hour, while others rambled. Some rotated in seconds, others much slower; some spun like ballerinas and some tumbled end over end like acrobats. Long before hitting the moon most asteroids had collided with other asteroids, perhaps many times, breaking apart or adding together.

Yet amid the chaos were patterns, signs of ordering forces at work. Most asteroids came in from the same general direction and came rotating in the same direction, motions hatched from the rotating gas cloud that had also generated the motions of the moon and the planets. There was a pattern in time: far more asteroids had joined the moon in its first billion years. Though few of the asteroids were close to round, and though some hit at low angles, most of their craters were close to round, some very close, for gravity and geology translated their impact energy into roundness, just as gravity and geology had shaped chaos into the roundness of the moon. There was the pattern of one lunar side being far more saturated with craters than the other, which held huge plains of lava.

When asteroids hit, the moon carefully measured their size, weight, velocity, and solidity by producing craters of different depths and sizes, and by producing shockwaves—very round waves—that spread through the moon and hit its magma core or its opposite side and echoed back, echoed many times, the moon ringing like a bell, yet ringing unheard. The asteroids fell onto previous craters and cut them in half or obliterated them and formed new craters or craters within craters. Asteroids smashed the rubble of previous asteroids into further rubble and dust, spraying it upward and outward for the thousandth time. The largest asteroids had hit so powerfully that they cracked the moon's surface and released its magma to flow into massive lakes, in which new asteroids made huge splashes and tsunamis and disappeared. Asteroids added their faint gravity to the moon's gravity and helped gather even more asteroids, grind even more dust, and bloom out of cosmic chaos even more roundness. The asteroids continued falling.

Yet the asteroid that was descending now was not behaving like any previous asteroid.

Some of the metal in this light had once been part of asteroids that were heading for the moon and that very nearly hit it; they would have forever merged their metal with the moon's chaos, stillness, and greyness. Yet as they were wandering toward the moon a stronger force had begun acting upon them, a much larger body not far away, diverting their course away from the moon, aiming and accelerating them toward itself. Then something odd began happening, something that had never happened to asteroids approaching the moon. The asteroids ran into resistance. They began heating up, melting away, igniting a fireball around them, a fiery tail behind them, igniting loud noises. Many of them broke apart into pieces and into ashes. Some of them endured long enough to hit the ground and punch out craters, but once again something odd began happening: this world began healing itself, salving away its scars, absorbing the broken asteroids into its own processes, washing them downstream, burying them, sucking them deep underground, melting them, combining them, launching them from volcanoes, raising them into mountains. By joining Earth, asteroids had joined an adventure far greater than that of the moon. They joined a world of action, ability, and color. They joined a world of water deep and endlessly flowing through hundreds of forms. They joined a world of air blowing through asteroid lips and giving them a name. They joined a world of fire forging cosmic swords into mountains. They joined an Earth kneading itself into ever richer earth. Instead of lying in unrelenting greyness, they were helping to uphold frequent rainbows. The asteroids helped propel Earth's forces and forms toward greater order, which one day would propel asteroids off the ground and into the sky, back into the space from which they had come, back into the realm of tumbling, unreformed rocks and metal, back to the moon they had nearly joined long ago.

This asteroid descended not with suicidal speed but with a slowness the moon had never seen before. It descended not with steady acceleration but with changing rates, moving slower as it neared the surface, actually pausing. It descended not in a straight line but moved sideways and changed direction slightly. This asteroid was not under the control of the moon's gravity but was defying it. Somehow this metal had escaped the force that had shaped all matter, and it had passed into the control of another force that had married itself to gravity so that it would not be bound by it. This asteroid was not predestined by its past momentum but had transcended the momentum of all matter since the Big Bang and taken control of its own actions. This asteroid was defying the fate of all asteroids. This asteroid had energy streaming out of it, not the energy of an asteroid burning up but a comet's tail streaming beneath it, and not a comet dissolving from the sun but a stream of energy carefully aimed and regulated, powered by its own internal sun. This asteroid was not dull grey but was bright gold and silver, was not modest with the sunlight but gleaming brightly, sparkling as it pivoted back and forth. This asteroid was not crudely shaped, not lumpy and pitted, but held many straight

lines, smooth surfaces, graceful curves, slender append-
ages, and symmetries small and large and exact—symmetries, symmetries, symmetries, symmetries that proved
this metal had left the universe of asteroids and moons
and fallen into some black hole of ardent geometries and
emerged reincarnated. It had enfolded itself with many
natural forces and laws and was now applying them to
move itself so skillfully. Yet in gaining form and ability, this
metal had lost a lot. This asteroid was not solid but hollow,
not strong but fragile, less likely to gouge out a deep crater
than to smash itself apart. This metal had traded endurance for configuration, immortality for talent. For eons
this metal had drifted along without caring about going
anywhere, without caring about being anything, without
caring about coming to an end, yet now it was moving
with extreme care.

And what was this? The asteroid held glass that was
very different from the tiny glass grains scattered across
the moon, which was fused by volcanoes and asteroid
impacts and solar flares. This asteroid's glass was smooth
and thin, clear and transparent, allowing the moonlight
to pass through it and the light inside to pass outward,
revealing the space within.

The asteroid's symmetries and forms and motions
were echoed by symmetries and forms and motions within. The glass showed two knobs that didn't look anything
like metal. The knobs held multiple contours and symmetries. They held multiple holes through which the outside world blended with the inside world, through which
various energies flowed: light and gas and waves in the
gas. Some holes remained open while others continually
opened and closed and reopened. The knobs moved for-
ward and backward and from side to side, but stayed close
to the window. Their motions were internally generated
motions, not commanded by gravity, electromagnetism,

27

planetary motions, geological forces, or weather. Their motions were the source of the motions of this former-asteroid metal. Their forms and symmetries had been translated into the metal's forms and symmetries. Their forms were too different from the formlessness of the moon and thus they had to continue moving to find a spot where they might fit in. From the craters and boulders, from the mountains and rilles, the moonlight rose up, carrying its images of chaos into a sky that had swallowed the moonlight without response for four billion years, but it turned out that something had been watching the moonlight for much of that time and now it had finally arisen and followed the moonlight home, followed it over the craters and rubble and dust, followed it along the jagged boundary between chaos and order, followed it to locate the different fates of moon and Earth.

The crafted metal skimmed over the craters and boulders, hesitating then accelerating, descending again, drifting to the right a little, casting a vague shadow that soon focused into a sharper shadow full of shapes and symmetries, overruling the crude shadows of boulders and rocks: Earth was eclipsing the moon; order was eclipsing desolation. The craft was searching for gentleness, a spot free of the violence of asteroid impacts, for it had been crafted by the same gentleness that wove molecules into DNA and into cells—no asteroid speeds were tolerated there. It had been crafted by the same gentleness that wove cells into corals and ferns and butterflies and eagles, and though it was true that this gentleness had started behaving violently, that eagles began grabbing fish and tearing them apart, even this violence was driven by the gentleness within that did not want to surrender to the chaos of asteroids. The moon continued trying to turn the craft into an asteroid and make it crash and gouge just another crater but the craft continued resisting, continued searching. It searched partly by broadcasting electrical waves at the moon, waves that bounced back up carrying messages about the surface's distance and roughness. This electricity was a magnification of the gentle electricity flowing through neurons. Moonrocks contained the same electrical forces and possibilities but had always kept them locked up deep inside.

The craft was also touching the lunar surface with its jet of energy, a fire new for the moon, Earth energy, a volcano pointing downward, skillfully levitating and moving the craft. This fire had been sparked by the metabolic fire within cells. It contained the torches by which animal murals were painted deep in caves; it contained the stained glass and mandala suns and moons that radiated the mystery of the cosmos. This fire brushed the moon with wonder. The moondust felt this Earth energy and began vibrating and jumping and streaming, not billowing upward in clouds, for there was no air to carry it, shape it, and smother it, but spraying outward fast and far. The moon dust learned something of what it was like to be Earth dust being driven by winds or tornadoes. This fire was digging a wholly new kind of crater.

When the craft was a few feet from the ground, its metal filament touched the moon and sent Earth electrons scurrying to turn on a little light within the craft, a new star in the sky, a star that contained four billion years of sunlight and cell filaments probing their surroundings.

The craft swayed slightly to balance itself, to complete the balance built into its four precisely measured legs. Yes, legs, another thing the moon had never encountered on asteroids, which never worried about landing upright. The craft's legs had grown out of flower stems and mighty trees, out of spider legs and heron legs and tiger legs, out of legs designed for crawling, walking, climbing, jumping, digging, and running. Legs were another story Earth would now tell the moon.

The engine finally shut off and the craft yielded to the moon's now-gentle embrace. The legs dropped onto the dust and pressed into it, forming four new craters, but the moon was impressed only physically and not consciously. The moon could not recognize what had happened. The flying dust settled back to the ground. Inside the craft, the knobs moved, moved closer to the window, and their holes moved. Earth was impressed by the moon.

Even from the distance of the moon, Earth had been visibly changing for four billion years. Its oceans, continents, ice sheets, and cloud content had grown and contracted and shifted positions and colors. Yet from the distance of the moon, from the blindness of the moon, the moon hadn't been able to tell what was happening on Earth.

Now Earth in concentrated form, all its continents and oceans and air, all its eons and evolution, all its forces and forms, had come to the moon. Two worlds that had been separated at birth were being reunited to tell their stories, a world of form and a world of chaos, a world of oceans and a world of dryness, a world of air and a world of nakedness, a world of sound and a world of silence, a world of color and a world of greyness, a world of deliberate flight and a world of shrapnel, a world of life and a world of death, a world of consciousness and a world of obliviousness.

From in front of rows of lights, lights lined up like no constellations yet born from them, two faces looked out on the moon, and in their eyes the moon was mirrored, finally mirrored not just in glass: the ancient moon, the chaotic moon, the cratered moon, the lonely moon. In their eyes the moon sparkled. The moon searched. The moon glowed with wonder.

2
OXYGEN

The lander sat quietly for awhile, gleaming with rejected sunlight yet still warming up from it. The windows showed motions within. Then two vents opened, one on top and one on the side, and gasses began flowing out. They hesitated, trying to discern how they fit into this new environment. They felt the moon's gravity trying to hold onto them, yet felt no atmosphere trying to include them, no winds giving them direction. They quickly decided they did not need to be here and began diffusing outward and rising upward, a few molecules straight toward Earth, as if they wanted to return to the place where they belonged.

Through a strange mitosis, Earth's atmosphere had generated a tiny baby atmosphere and sent it floating away, to the moon.

This dispersing air was the silent breath of clover, and the roar of dinosaurs. It was the waste product of forests, and the gift of birdsong. It was the violence of thunderstorms, and the grace of butterfly wings. It had been generated by redwood trees hidden in fog, by seaweed bobbing with lunar energy, by desert cacti and tropical jungles and arctic lichen. It had been breathed by moths searching for moonflowers, by fish riding the tides, by birds migrating by starlight, by cats prowling by moonlight, by singers singing about the moon. It had been purified by the same sunlight that was falling onto the moon but making only heat, setting molecules vibrating but not building anything. It was ancient, and it was always new. It had been circulated around Earth and in and out of rock strata for billions of years. To endless creatures, it had granted life and growth and motion and voice. Even as the air was transformed back and forth, it had transformed life into ever new forms. Even as life changed form, the fire within it remained the same and was passed onward. Inside life's mazes, air was far more powerful than it ever was as hurricanes.

From all of this history and activity, the oxygen had been removed, and now it was dumped onto a landscape where it was homeless and purposeless. Yet the oxygen and the carbon it was carrying into the sky had finished their long adventures on Earth by steering this craft to the moon.

The oxygen was flowing over moonrocks rich with oxygen, some 42% oxygen, not much less than the oxygen content of Earth's crust, but all of the moon's oxygen was imprisoned inside rocks. Any oxygen freed by meteor impacts or volcanoes had quickly left the moon and gone wandering in space, a bit of it to Earth, where it discovered the freedom and camaraderie of an atmosphere, of flowing for thousands of miles, of drawing patterns and carrying water, of giving a voice to a world, of meeting itself at the intersection of eyes and rainbows.

On the side of the lander a door opened, and more oxygen and carbon dioxide puffed outward and spread upward, some toward the breathtaking Earth.

More oxygen appeared in the doorway, not as escaping gasses but as dense concentrations inside a spacesuit and metal canisters, connected by pipes, tubes, wires, fans, and filters. This atmosphere held its own weather system, with high pressure flowing to low, with moisture condensing. The air flow was not steady but fluctuated with motion and excitement, the excitement of Earth oxygen meeting its cousin oxygen. The spacesuit air embodied the half billion years in which life had held the oxygen content of Earth's atmosphere at 21%, more than a hundred times the oxygen content of Mars air. If that 21% had dropped to 17%, animals would have been unable to breathe, and it if had risen by a few percentage points, organic matter would have been dangerously inflammable. For eons, life had gone to great trouble to avoid oxygen, yet eventually some adventuresome cells had noticed that oxygen's toxicity meant energy, and they figured out how to use oxygen to construct and move molecules and to move entire animals around. Eventually, cells decided that life arose from the gods breathing into clay, and that human breath held the soul. Those cells had remained adventuresome, and now they had ridden the fugitive wind to the moon.

The astronaut breathed deeply, deep into time, deep into Earth, deep into life, breathed with the lungs of whales, the throats of giraffes, the mouths of Rex, the noses of elephants, the voices of wrens, tapping the holy, cosmic winds that had given life to the gods themselves.

3
THE LIGHT

Amid the greyness in which the moon had dressed the sunlight for billions of years, the lander gleamed silver and gold. On the lander's shadow side, light appeared and widened. A door was opening, the door between moon and Earth, between chaos and form, a time portal between the beginning and the arrival. Fully open, the door was a square of light, the combined light of a constellation of lights of various sizes, colors, and purposes. All this light was transfigured sunlight, the sunlight that had flowed onto a once-dull Earth and impregnated it with energy, form, and color, making flow the oceans, storms, and rivers, making turn the seasons, fusing atoms into molecules, molecules into cells, and cells into redwood trees and whales. This light had sparked the creation of billions of life forms that filled and transformed Earth. Sunlight became the flashes of fireflies, the trotting of zebras, the songs of larks. It pumped blood through veins and vast migrations through the skies, seas, and land. It painted Earth with colors far more vivid and diverse than rainbows. It masoned deep layers of rock and memory and energy. It flowed into humans and became comforting campfires and brilliant cities, became wheels turning and mouths speaking thousands of languages, became telescopes and church murals asking the origins and mysteries of light. It assembled in fuel tanks and batteries and computers and propelled itself to the moon.

The sunlight glowing from the doorway was eclipsed by shapes moving within, shapes projecting vague shadows onto the ground. Transfigured sunlight, with sparkling eyes, gazed into the deep, dark blindness of the moon.

4
BALANCE

From the beginning, from too much chaos, the solar system was seeking balance. From a primordial cloud frantic with collisions, planets and moons emerged, and their gravity helped further sort the chaos into stable shapes and orbits. Yet chaos nearly triumphed over the baby Earth when a small, wandering planet collided with it, shattering both. As their debris rebalanced itself, it formed a new Earth and a moon with a stable if slowly receding orbit. Out of the balanced motions of Earth and moon there eventually emerged a further motion, a spacecraft flying with astronomical precision from Earth to the moon and into orbit around it—now Earth was orbiting the moon. As the spacecraft neared the lunar surface, it broke loose from gravity and began asserting its own motions, wobbling, trying to stay balanced, trying to find a place to balance itself on the ground. And now, as the astronaut climbed out the door backwards and onto the ladder to the moon, he wobbled again, unsure of the low gravity and of his footing and handholds.

The astronaut's motion was a continuation of Earth's ancient and utterly reliable motion, so why, after Earth had generated billions of years of further order, was matter now struggling with its motion and balance?

As order had emerged into all the orders within life, it diverged ever farther from astronomical order and opened up new possibilities for chaos: disease, accidents, predation, and aging. For life to move, it needed systems for overcoming gravity and evading a world of obstacles. It needed limbs of many kinds, fins and wings and legs, of many shapes and sizes and compositions. It needed eyes and balance sensors and brains to coordinate motion. The astronaut was packed with eons of experimenting with and learning about motion. He contained the blind wriggling of cells in the ocean mud, and the rising of fish to the morning sun. His motions seemed to remember the awkwardness of the first fishes to happen onto land. His motions contained the leaping of frogs, the creeping of lizards, the tunneling of worms, the flapping of birds. He contained stealth and charges, stampedes and migrations. His training had been done by mice-size mammals climbing into trees and by primates climbing down—his ladder was a symbolic tree. His grip had been felt by spears and bones, paintbrushes and funeral offerings. His bones had been reinforced by ages of falls, breaks, and deaths, by ages of conflict between order and chaos.

Rung by rung, step by step, hand by hand, breath by breath, Earth flowed onward, tilting and hesitating, relearning its partnership with the moon. Very soon, Earth would know how to step lightly upon the moon.

5
FOOTSTEPS

The dust had sat there for a million years, or maybe a hundred million years, motionless, waiting for something to happen, or for nothing to happen. The sun came and warmed the dust and colored it grey, and then the sun left and the dust bled its warmth and color away, over and over again. Occasionally the dust vibrated from a distant asteroid impact, but it had been a very long time since an asteroid had hit close enough to reshuffle this dust or add new dust to it. Once, this dust had been solid rock, the rock of the lunar crust or lava flows or asteroids, but asteroid impacts had smashed the rock into fragments smaller and smaller, into boulders and pebbles and dust. The sun and the night, the heat and the cold, joined together to crack the rock toward dust. Asteroid impacts tossed the dust back and forth, burying it for a thousand or a million years, excavating it, then burying it for another million years. The identities of millions of asteroids, their origins and shapes and compositions and wanderings, merged into the story of the moon. In its molecules the dust had not forgotten the asteroids it had been, but it had entirely forgotten the energy it once possessed and the violence it had inflicted upon the moon, and now it sat powerless to move itself. Obeying gravity, it still tried to fall to the center of the moon but was trafficjammed by all the rock beneath it. The dust added its gravity to luring more asteroids to the moon. As the era of heavy asteroid bombardment had faded away, the dust sat motionless longer and longer, counting not hourglass time but the timelessness only the moon could perceive. The dust might be ten or twenty inches deep, but there were variations from spot to spot, ridges and bumps and valleys of dust. The dust was the crowded graveyard of particles from the solar wind, solar flares, and cosmic rays. Occasionally a micrometeorite, a speck of rock, punched a tiny crater into the dust.

And then one day, without any premonition, a strange new crater appeared, a shape the moon had never seen before. Compared with the craters all around it, it was small and shallow. It was not round, but oblong. It did not have a central pit and slopes to a peaking rim, but was even from end to end, with ridges and groves running from side to side, ridges and grooves with very straight lines. The crater's shape curved steadily. In billions of tries, asteroids had never produced such geometry. This crater was not gouged violently, but impressed gently. The moon resisted asteroids, swatting them apart, answering chaos with chaos, but this crater it accepted calmly. This crater was not the endpoint of a long and ancient trajectory but had originated right here, right now. Asteroids had never cared where they fell, but this crater spot had been selected carefully. A wholly new force was loose upon the moon.

Another crater appeared, then another, and another. This asteroid had remained intact and was still at work. The craters pointed a direction, not the oblivious trajectory of an asteroid but a deliberate course, which revealed the larger course on which the asteroids, the moons, the planets, the stars, and all the galaxies had been flying all along.

Another crater appeared, but this time its maker remained in place, snug against the dust. The dust and this asteroid might be touching physically, but in other ways they were far apart, in parallel universes. They were made of the same elements, but their elements had gone on very different journeys.

The elements inside the dust were a muddle, random

mixtures blurred together, seldom a grain of a pure element. The confused elements added up into larger confused shapes, into ridges and slopes quickly contradicting one another, into bumps and holes and debris. The confusion added up into greyness, blurred variations of greyness. The dust sat motionless, except on the inner level where electrons swirled around nuclei, but this atomic eagerness to make bonds and build forms had gotten no farther than grey dust.

Atop the dust now stood a tower of order, of form upon form, substance upon substance, activity upon activity, ability upon ability, timing upon timing. Its elements had been carefully sorted, concentrated, shaped, and assigned duties. Its elements formed smaller shapes that added up into larger shapes, simpler shapes that added up into more complex shapes, simpler talents that added up into genius. Atoms formed molecules, molecules formed units of molecules, units formed communities, and communities did different tasks that added together into a harmonious whole. The tower was a maze of shapes and chambers and tunnels, an architecture of both strength and flexibility. The tower was filled with motions, many actors and sizes and layers of motion, solids and liquids and air and electricity in motion, flowing into one another with perfect strength and timing. Elements and molecules that were locked up within the moondust, or that barely existed on the moon, were free and flourishing here, energized and energizing, shaped and shaping. Upon the dry moon a river was flowing; upon the airless moon a breeze was blowing; upon the oblivious moon an idea was guiding. All the motions inside the tower merged to make the tower itself move, to make further steps.

The moon experienced the footsteps only with dust impressions. The moon could not recognize that it was being visited by another planet, that its own destiny was not the only destination, and that out of the formlessness from which both had begun, Earth had arisen into form and life and consciousness.

The footprints were shaped like cells, like paramecium, like trilobites. The footsteps were being made not by one human alone but by the whole long procession of life, all still alive within him. His heartbeat had been passed from life to life for billions of years. His breath had been resuscitated by millions of species. His metabolism still contained the swirlings of the first cells and of the primordial sea. His nerve traffic had been groomed and improved by countless creatures watching the seaweed swaying, watching fruits and flowers blooming, watching for shadows moving, watching the sunsets. His muscles had been toned by countless creatures walking, running, jumping, handling. His footprints were a continuation of footprints in the Burgess Shale and the Laetoli volcanic ash. His footprints were being made by bacteria and fish, lizards and centipedes, dragonflies and eagles, frogs and dinosaurs, sloths and cheetahs, kangaroos and monkeys. Humans contained everything except the memory of all they contained, and thus they could imagine their footsteps to be merely a giant leap for themselves.

Earth was showing the moon what it had been doing, all it had been learning for four billion years. Earth was giving the moon a gift: the shapes of cells, the symmetries of life, the geometries of crops and Greek temples. The moon had been impressed only by violence, but how it was being painted by an impressionist.

Shortly before coming to the moon, the astronaut making these footprints had walked along the ocean beach a few miles from the rocket launch towers. He had turned the sand into footprints. But his footprints were soon erased, erased by the moon and the waters it pushed high onto shore, for on the hyperactive Earth even the inert moon gained superpowers of action. The sand that had formed his footprints was still there, smooth and rolling, with no memories of the sand castles humans had built or the journeys that were so important to them.

Now the same feet were making footprints on the lunar beach, footprints that might last a million years, or more. The moon would honor human shapes long after their own skeletons had given up and their species had forgotten them. Their footprints would last longer than meteor craters on Earth. Yet gradually their footprints would decay, yielding to micrometeorites, gravity, moonquakes, heat and cold, and new dust arriving from space, yielding to the tides of the cosmos.

The astronaut turned and looked back at his footprints, at the way he had come. Beyond his footprints he saw craters, small and large, all the way to the horizon. The craters too were footprints, the footprints of asteroids, marking the long journeys they had made through space and time. In the strange journey of the cosmos, asteroid footprints had transformed into human footprints.

The astronaut stood there looking back, looking into the formless dust become form, seeing the strangeness of the moon, seeing the braille craters all the way to the horizon, seeing deep into space and time, to the birth of the solar system, seeing deep into the chaos from which life had arisen.

6
THE ROCK

The rock had sat there for a hundred million years, or maybe three billion years, motionless, fiercely loyal to gravity. In size and shape and color the rock was barely different from thousands of rocks around it. Very slowly the rocks were shedding a bit of themselves into dust.

And then one day this rock, only this rock out of thousands of similar rocks nearby, felt a strange force acting upon it, summoning it.

The rock rose from the ground, and hovered there.

The rock felt gravity clinging to it and tried to obey, but gravity was being overruled by something stronger. Not volcanic force, not a meteor impact, not a moonquake. In anti-gravity, the rock hovered.

From the metal hand that had picked it up, the rock was passed into a softer hand, which folded around it. Two worlds were shaking hands, a world that had become alive and a world that had remained dead. They had known each other long ago, when both were clouds of gas and rock, when both were starting from the same place, but they had gone on different journeys and arrived at different fates. Now they were meeting again and comparing their faces and their stories.

The rock was hard and rough, while the hand was soft and smooth. The rock grew hot or cold at the mercy of the sun and night, while the hand maintained a steady temperature. The rock was inflexible, while the hand moved and flexed. The rock was oblivious of the hand, while the hand could feel the rock's density and shape and feel its own pulse against the rock. The rock was naked, while the hand was mummified to defend itself against the moon, against death.

The astronaut looked at the moon in his hand, at the moon glowing. The glow lit up memories in a strata the moon did not have, in the folds of the unconscious, memories of a full moon rising over the Mojave desert, over the mountains and alluvial slopes and sand dunes, implying the existence of creosote bushes and wildflowers and coyotes, making the harsh desert gentle and beautiful, making Earth seem so strange, as strange as it genuinely was.

7
HYDROGENESIS

The hand was soft and flexible because it was made of water, of two elements that liked to flow, hydrogen and oxygen. In the sun, hydrogen flowed in massive currents, circling deep and arcing into space. Around Earth, oxygen formed a shell of motion. Together, hydrogen and oxygen formed the water that flowed continually from seas to clouds to rains to rivers and back to seas. The astronaut's motion was generated by the energy flowing from the sun, the energy flowing out of oxygen, and the restlessness of water. The astronaut was walking upon hydrogen and oxygen inside moonrocks, but they were locked up, being punished for being in the wrong place at the wrong time.

The hydrogen in the astronaut and the hydrogen in the moonrocks had known each other a long time and gone on great adventures together. They had been thoroughly mixed together in the vast cloud of gas and dust from which the sun and solar system were born, and they easily could have been sorted differently and ended up in other bodies, the astronaut hydrogen in moonrocks and the walked-upon hydrogen in the walker, or both of them burning in the sun. Hydrogen karma had decided. Before the solar system, this hydrogen had drifted through space in thin but thickening clouds, and before that it had been part of other stars but escaped being fused into heavier elements and been shrugged back into space when the stars died. Perhaps the astronaut hydrogen and the moonrock hydrogen had participated in the same star, knocking into one another and ricocheting away for a few billion years. Long before that, this hydrogen had emerged together from the Big Bang, swarmed madly together, been shot in the same direction together, helped form the same galaxy, and been funneled into the same section of that galaxy.

When two hydrogen atoms collided in the Big Bang, they could not imagine that some fourteen billion years later they would meet again, one as part of the moon and one as part of an astronaut walking on the moon. The universe was launched on a strange, improbable, impossible journey into order, into solidity, into life, into consciousness. The ignition of the moon rocket, its roiling flame, was a faint echo of the Big Bang. The journey of humans to the moon was but a continuation of the long journey of matter out of formlessness. The lander was a time machine taking the elements to their past, letting them see where they had come from and what they might have been, giving them a mountaintop view of what they had become. The entire universe could have been nothing but worlds of craters and greyness, nothing but hydrogen clouds without a spark of light, nothing but emptiness forever.

8
CALCIUM

The moon's deadness was outlined in chalk, in calcium, not calcium processed into thin lines or distinct shapes but calcium blurred thinly with other elements, as if the calcium had already bled from a dead body, as if even its bones had turned to dust and blown away. The moon had been murdered, and lime had been tossed over its grave.

The moon was an impurity that was sealed off with calcium, becoming a bright pearl only impersonating perfection.

The moon was a stalactite, calcium leaked from the rock of the ultimate cave, congealed into irregular blobs, an hourglass counting in eons, a dark, deep secret.

The moon was a calcium cliff, a tall white cliff of Dover, holding invaders at sea.

Yet the invaders had come: calcium had emerged from the calcium sea where it had built coral castles and stormed them; calcium had emerged from the calcium caves where it had spun castles not to serve anything or be observed by anyone but for the oblivious universe's own sense of sheer fun; calcium had emerged from seashells whose lines spiraled outward and outward in perfect ratios until they intersected with the orbit of the moon; calcium had emerged from marble quarries from which it became rows of Greek gods and goddesses upholding the order of the cosmos, including the moon; calcium had emerged from the teeth of the dragons and wolves that had swallowed the waning moon; calcium had emerged from theaters in which the limelight had cast light and shadows into the darkest and brightest mysteries of human existence; calcium had emerged from the river of milk that nurtured the life that screamed for living and growth and that imagined the whole sky was a river of milk flowing into life and buoying it upward; calcium had emerged from the primordial golden cosmic egg.

Earth calcium had come to moon calcium and was impressing it with a miracle the moon could never perform: a simple footstep, the artwork of calcium, of a vast calcium architecture moving with the authority of a billion years, with the strength of coral reefs and the grace of an egg's curvature and the agility of a lion's run, moving reliably even here far outside its home realm. Earth calcium pressed moon calcium a tiny way toward greater concentration. It was a giant leap for calcium.

9
TIDES

The ocean tried to reach the moon.

As the moon passed over the ocean, the waters felt the moon's strength and responded to it, rising a dozen feet high, raising a bulge of water thousands of miles wide, raising seaweed and fish towards the moon like an offering. As the moon passed onward, the massive bulge of water followed it for thousands of miles, trying to catch up, joining the moon's orbit, revealing a loyalty to something larger than water and Earth, revealing its membership in the cycles and energies of the cosmos. Even in the ocean depths, so far down that the moonlight was smothered into blackness, the moon's power was still stirring the sands and steering the fish. The ocean's bulge would have followed the moon forever, around and around Earth in a geometry of motion as perfect as the solar system, but Earth's perfection was marred by continents with all their irregular shapes, and eventually the bulge ran into land. Still trying to climb to the moon, the waters now climbed onto land, rising many feet high, rising for hours, breaking into waves flashing with moonlight, rolling sand grains sparkling like tiny moons. As the tide rose, it erased the sand lines left by previous tides and lifted driftwood and shells and bones and carried them further up the beach, toward the moon. Every wave drew a new wavy line on the beach. The moon had inserted its invisible hand into the ocean and turned it into a paintbrush for making pictures and collages on the beach. Over the millennia the tides redrew the shapes of entire beaches, piling them higher or tearing them down, making them plumper or skinnier, moving them backwards or forwards, changing their curves, erasing entire islands and peninsulas and rebuilding them elsewhere. The tides' comings and goings had been imprinted in sand and then rock a billion years deep. The moon also caused tides in the air, encouraging the winds. It caused tides in the earth, stresses that might encourage earthquakes. Over the eons the moon's tidal friction had stabilized Earth's rotation, slowing it down, lengthening Earth's days, preparing a viable chronos for life. And perhaps some four billion years ago, in warm tidal pools confused with molecules, the tides had helped organize and energize life. Today the moon shows life hatching and swarming and living. The tides revealed the mystery of gravity, occult action at a distance, the sensuous curving body of space-time, matter's appreciation for itself, the force that holds together and moves the entire universe, reliable to the second and to the inch yet still inexplicable, still unable to keep the universe together and running forever.

Tonight, the tide might be slightly different than it was a few nights before, if different only by a few grains of sand, for a few days ago Earth and moon had readjusted their weights. Nearly 100,000 pounds of Earth had gone to the moon and become a moon of the moon, and some 33,000 pounds of that had landed on the moon, some of it as propelled gasses, most of it in one solid body, including some 350 pounds in two living bodies. The moon had summoned her children. Life was now contributing to the moon's draw upon the oceans and the beaches and the sea turtles.

This 33,000 pounds of Earth matter had long influenced the moon, and more strongly than the same amount of moon matter had influenced Earth. It had helped cause underground tides, stresses, cracks, moonquakes, and surface landslides. Looking out their windows, the astronauts saw dust slopes they had helped to draw. This Earth matter had helped to speed up the moon's rotation and nudge the moon ever farther from Earth, requiring Earth to now spend more energy and time to cross the gap it had made.

The astronaut bodies were mostly water, water that for four billion years had been feeling the power of the moon and performing the tides. Their eyes sparkled in the same way they had sparkled beneath the ancient moon, a much closer and larger moon. Their water had flowed all over Earth, flowed on the ocean surface and miles underwater, flowed as momentary waves and as currents thousands of miles long, flowed in patterns changing as the continents changed, flowed into the air and into clouds and over the land and back down as rain, flowed as rainbows, flowed as snow shimmering with moonlight, flowed as glaciers prisming the moonlight, flowed as mountain cascades and waterfalls, flowed as rivers eroding mountains into plains and plains into canyons, flowed underground and made caverns and mineral hourglasses that never knew moonlight, flowed back into the ocean for the millionth time. This water had flowed over an Earth without life and helped create it and photo-fuel it. Formless water had flowed into bodies and taken on numerous forms and yet never belonged to any of them and flowed back out again, its restlessness carrying life back into formlessness and yet onward into new bodies and new species. Formless water had merged with forms that had given it new identities and abilities, the forms of cells and corals and fins, leaves and roots, wings and legs, fur and scales, faces with which to disguise itself, brains with which to shimmer with moonlight and wonder.

Now the ocean had come to the moon and was flowing upon it. This was odd, for water loves itself and was always trying to return to itself. The sun was always trying to separate water from itself, hauling it into the sky and over continents, dumping it thousands of miles from home, but water was determined to return to the ocean, even if it had to endure many obstacles and transformations along the way. In human form, water was drawn to itself in new ways. Water flocked to seashores and felt soothed and excited by the surf. Water went swimming, sailing, scuba diving, and surfing. Water paid a lot of extra money for houses overlooking seas and lakes. Water felt it romantic to go on ocean cruises and to honeymoon at famous waterfalls. Water loved to paint pictures and write poetry about itself.

But now water had chosen to leave itself far behind. Now water looked up and saw a single drop of water in the sky and felt lonely, felt the sun trying to suck it out to where it could never return home. Water was flowing across a deep, unimaginable dryness, beaches without seas, an inertness utterly different from itself. The ocean hadn't stopped moving for billions of years, while moon boulders had sat frozen. The ocean had whistled and roared while the moon was locked in silence. The ocean had flowed in ever more elaborate patterns, in tiny and precise tides of molecules, in ocean-wide tides of birth and migration, and then the ocean had encapsulated itself to move onto land, the too-dry land that required new shapes and strategies for drawing in water and replenishing its flow. Now the ocean had dressed itself in spacesuits for going to other lands where it did not belong. The ocean echoing inside white seashell helmets was not really their own heartbeats but the ancient pulse of the ocean. The salt in their sweat contained eons of hard work. Their motion was being propelled by endless rivers and tides, by hurricanes become rockets. The footprints they left in the dryness had been practiced by eons of waves washing ashore, by raindrops leaving tiny craters in desert sands.

The moon was not as inert as it seemed. It had a secret identity, a split personality, for with its invisible gravity hand it stirred in water abundant motion and power and change. The tides shimmering with images of the moon truly were the moon's mask. The moon walked all over Earth's beaches, one giant leap for moonkind. The moon reshaped beaches into its own smile. The moon had left its signature in DNA. Through the fish it herded, the moon wrote love letters to Earth; through dolphin minds, the moon recognized and enjoyed its own image and vectors; through children building sandcastles and sand rockets, the moon discussed the ultimate powers of the universe. Through its agents on Earth, the moon expressed the dreams that lay locked inside grey, motionless, silent rocks.

The water in the astronauts' bodies already knew the moon well. It has obeyed and portrayed the moon for eons. The moon had summoned it onto beaches, splashed it against cliffs, squeezed it through coral reefs, taught it to articulate the songs of the moon. Every atom of their water had been thoroughly trained in how to respond to the moon's gravity. Now this water was learning the moon anew, from much closer, feeling it pull their spacecraft downward, feeling the tug-of-war between spacecraft and moon, feeling the moon wobble them down the ladder, feeling the moon not as a flowing force but as something solid beneath boots, feeling their baby steps on the moon. The ancient tide rolled along on a new beach. The river Ganges, reincarnated, flowed immortally. Water had flowed way uphill. The moon had been tugging on this water for eons, tugging it away from Earth and towards itself, and finally the water had come, finally it was standing here at the altar of its tombstone god. This water had gravitated towards the moon through a new type of gravity, a force generated by brains, in which the moon stirred not just waves but brainwaves, not just motions but emotions, not just tides but good tidings. The ocean that had always sympathized with the orbits of the cosmos had come to find out exactly how it fit in.

10
TIME

The moon might seem to be timeless, if time is defined by all the processes and changes found on Earth. On the moon, rocks billions of years old sit patiently waiting for another billion years to pass, while Earth rocks born in the same era were long ago eroded away and moved around and transformed into new rocks, many times over. A lunar landscape today might be indistinguishable from the same scene millions of years ago. Yet time on the moon passes at the same rate as it does anywhere else. The moon has never confused time with events or changes—only confused planets do that. An hour of absolutely nothing happening is still an hour rich with time, time unworried, time unhurried, time running deep.

Yet the moon does have its own ways of registering the passage of time. Every boulder and pebble is a sundial casting a shadow that grows shorter and longer and circles through the long lunar days. The rocks also dial the sun by warming up and cooling off, and by gradually cracking and defoliating into dust. The moon counts time with impact craters, the gear teeth of the rolling solar system, and with space dust falling and filling those craters back in. Far better than Earth, the moon remembers the history of the solar system, its more intense asteroid bombardments. The moon counts time with its steady loss of internal heat and lava flows. The moon counts time with the ticking of its isotopes and with its gradual drift away from Earth. The moon's rotation and orbit around Earth makes its timeless face a giant and precise clock.

Yet in the distance, on the far more volatile surface of Earth, the moon began registering time in far more intricate ways. The moon guided the spawning and hatching and migrations of many species. Through the hands of an animal especially haunted by time, the moon began carving images of itself onto bones and rocks, images of crescents growing into circles and shrinking into crescents again and then invisibility, in cycles of twenty-nine days. Living bones carved the moon onto dead bones that might symbolize both endurance and impermanence. By torchlight deep within caves, hands painted murals with moons. The moon inspired Earth to rearrange itself, to haul heavy stones for miles and raise them high and in precise alignments to mark the moon's pathways and cycles, to allow life to stand within this cosmic portal and align itself with the moon and cosmos.

And then one day, one very long and patient lunar day, time on the moon began rushing. Time became a circular cage in which time circled like a trapped animal. Time was so restless and confused that it needed three hands to mark three layers of time, three hands of different sizes turning at different rates, all turning in the same direction, declaring that time had a direction and segments short and long and numbered. Time could be "watched" with a little machine that gave the illusion of control but that actually portholed an ocean of time relentless and beyond control. All these hands were attached to a much thicker hand that was also moving constantly, sometimes in circles but mainly back and forth and up and down, picking up things, moving things, connecting things, collecting things, moving as quickly as possible, competing against time, watching the time, being reminded of the time by a planet that considered a two-second radio delay to be wasted time, running out of time, trying to fit as much activity as possible into an hour.

The watch was the metal embodiment of a cell, of biological circles and schedules. Time had gotten itself entangled with a body, with heartbeats and breaths, a body full of echoes and mirrors in which time was amplified into an obsession, for lives were placed in charge of only one drop of oceanic time and had to make the most of it, defending it from time hunger and time decay, extending it as far as possible, with constant fear of running out of time. Through life's eyes time saw itself as birth and growth, sleep and awakening, eating and hunger, mating and migrating, sunrise and sunset, seasons and years, aging and death and decay. Time became feelings: yearning and worry, joy and regret. For a stimulus-craving animal, time became excitement and boredom. Time became memories and hopes and plans. In dreams, time revealed its true strangeness. Time became wistful songs and celebratory poetry, time-haunted novels and plays and films, time-fighting sculptures and time-melting paintings. In rivers, time saw metaphors for itself. Time became a giver and a thief, sacred and satanic. Silent and invisible time invented a thousand ways to make itself articulate and visible, including calendars set by the moon. Time dug into its past in ruins, fossils, rocks, and starlight. Time tried to beat itself with cosmetics, Olympic races, and tombstones. Dying time clung painfully and desperately to time, but always died.

On the moon, living time confronted unlived time, deep, primordial, undigested time, closer to pure vacuum time free of any changes. The relativity twins Earth and Moon were reunited after long journeys to find they had aged far differently. The time-traveling astronauts lived time not just with their watches but with gauges showing their declining air supplies, where zero would mean the end of time. They lived time with thirst, with declining water levels in bodies and water reservoirs. They lived time with heartbeats that speeded up with their motions. They lived time with sweat, with muscles flexing and sore, with tiredness, with activities scheduled to the minute. They hurried from spot to spot and from task to task, in another dimension than the moon, mayflies on an elephant, puppies in a Zen garden, knights of progress on a world decaying, never standing still for an hour to try to glimpse the moon's own pace, though this might require a thousand hours. Their frantic metabolism impacted the moon, stirring the dust, moving rocks, casting shadows—but only for a moment, only shadows that would soon leave the moon to its own shadows. Beneath their feet, the moon's isotopes ticked patiently, never noticing them, never grasping their urgency of time.

The astronauts looked at a cratered plain of unlived time and found it strange: empty, ominous, somewhat threatening, yet utterly fascinating. With their pitifully short lives they had come to try to touch deep time, to find it locked within rocks, to grasp it not just with hands but with minds, to take it home, to open rocks and release their time travel stories, to let time touch their lives.

Before lifting off, the astronauts ejected onto the lunar surface the boots they had worn for their moonwalk. The boots would lay there for thousands of years, casting shadows alongside rock shadows, surrounded by the footprints they had made, footprints the moon might guard for millions of years, its memory of the stream of time. But the feet that had made those footprints would soon walk along a beach and make footprints that would quickly be erased by the moon-driven tide, and before long, those feet too would be erased, erased except perhaps for some bones like the bones on which longing hands had once carved images of the waning moon.

11
THE PRISM

Relentlessly, religiously, like a monk in self-denial, the moon was grey. The volcanic robes were grey, the craters and boulders and dust were grey, the hills and valleys and plains were grey. The greyness did have variations, darker or lighter tones here or there, and up close the rocks might show flecks of white and black, but together the flecks blurred into grey. As the sun changed angles through the long lunar days, the moon might change its tones, but they remained loyally grey. The sunlight was rich with hidden rainbows and always offering to paint them onto the moon, but the moon always refused.

Yet suddenly the rainbow had appeared in the sky and arced downward and touched the moon. Now the rainbow glowed amid the greyness. On its outside the rainbow preferred gold and silver, in many variations, and on its inside the rainbow held its full spectrum of colors, colors the moon had never seen before, colors with different shapes, including round colors that could see the moon's greyness. The rainbow was inclined to see the moon as dreary, "colorless." Soon the rainbow opened and flowed out, tentacles probing the greyness.

Earth was a crystal ball, a prism that had fanned the sunlight into rainbows, painting its once grey self with thousands of variations of color, millions of shapes defined by their colors. Earth became red lava and autumn leaves, blue waterfalls and birds, green fields and frogs, orange corals and fruits, yellow rocks and wildflowers and bees. Earth broadcast an endless rainbow as the sunset swept ever onward, and sporadic rainbows with the rains. Earth partitioned itself into eyes of many colors for recognizing colors. Earth handed itself hands that loved to arrange colors into patterns. Altogether, Earth's colors added up into a blue sphere streaked with white, green, red, tan, and brown.

Earth had shined its colors onto the moon for eons, but the moon had been unable to see them or dream of turning colorful. Now the moon did its best to turn the astronauts grey, finding a foothold on their grey ambassadorial boots and climbing up their suits, clinging and biting. The moon filled their eyes with greyness, turned their color film into black and white, and made greyer the reflective lander. Yet the moon only reminded humans that they belonged to the rainbow.

12
GOLD

Embedded in asteroids chaotically shaped and aimed, gold had fallen from the sky. When the gold hit the moon, it scattered itself over the landscape, mixing itself with many other elements. Occasionally, new asteroid impacts relaunched it and reshuffled it. Then it sat undisturbed for eons, vaguely sparkling in the sunlight and Earthlight.

And then one day gold floated gently through the sky, sparkling, faintly and then brighter, tiny and then larger, shapeless and then a sculpture, flickering as it pivoted slightly. It came: a banner of gold, more gold than was contained in dozens of square miles of the moon, gold glowing with sunlight, deflecting the sunlight and defending the brazen shyness within.

Asteroids sprinkled with gold had also fallen onto Earth, but here gold was reworked by rainfall and rivers and sedimentation, by tectonics and volcanoes, concentrated into lumps and strands, and poked through the surface for animals to smell and taste and find to be worthless. Moon gold had never been smelled or tasted or judged for its usefulness in any other realm; it had sat unnoticed and unnoticing, content to be itself, content to reflect the sun a bit more truly than did lava.

Even many humans had ignored gold as too soft a metal, but out of their genius or perverseness others had decided it was pretty and thus had value. It was pretty because of the way it reflected the sunlight; its very name came from "glow" or "gleam". Because gold would reflect the 300 degree Fahrenheit sunlight of space, it was coated onto foil and wrapped around the lower section of the lunar lander to prevent it from overheating. It defined the landing as a thing of beauty and value. The most malleable of metals, gold could be spread so thin that it became transparent, and thus it was coated onto the helmet visors that would protect air-born eyes from the glare of the naked sun. The astronauts would see the moon and Earth tinted with the beauty and value of gold.

Considering how often gold has been transported and transformed into new forms, it was possible that a bit of the gold on the lander had once been part of a mask that gave a pharaoh an imperishable face for the afterlife, only for it to be looted and melted down for the sake of the living, for the sake of the greed that had never touched the moon. The gold could have been part of a Greek statue of Apollo the sun god or Selene the moon goddess, paying tribute to the mysterious powers that gave order to the cosmos and gave life to humans. It could have been part of an Aztec altar that required blood to keep the cruel cosmos circling safely, an altar that inspired Europeans to explore and bloody the Americas, an altar turned into a Spanish altar that offered the redeeming blood of a cosmos created in benevolence. It could have been part of a Chinese necklace that bragged that a woman had beauty and value and that helped pass her genes and face into far future generations. It could have helped fill a tooth crater and stop the biological decay the moon had never known. It could have been coins that paid for a birthday or a funeral. It was part of the vast and terrible complications of matter that had broken out on Earth.

While moon gold sat scattered in a land of unattachment, Earth gold had become moon-round rings that embodied the formative powers that had bound atoms into crystals and cells and trees and pharaoh's masks and that bound two bodies into staying together and creating more bodies, rings on the fingers that would land upon the moon and pick up rocks that assayed the value of Earth.

13
ALUMINUM

The lander glowed with sunlight, glowed differently from surfaces with different materials and colors and textures, some flat and some round, some smooth and some crinkled, some intended just to reflect the sunlight. In places, in angles, the lander gleamed. The lander was made mainly of two elements, aluminum and titanium, which were also present in the lunar soil, aluminum in large amounts, in proportions close to those found on Earth. But the lunar aluminum did not set the moon glowing, at least not much.

Once, five or seven or nine billion years ago, the aluminum and titanium of both Earth and moon had generated a far more intense gleam. It had swarmed inside the same stars, maybe hundreds of stars, that forged it out of lighter elements and sent tiny birth announcements flying into space. The light once generated by the atoms of the lander were now five or seven or nine billion light-years away, glowing onto many moons and perhaps a few lunar landers. The stars began sorting out the aluminum and titanium, but only by their weights and gravities, not by whether they would end up in Earth or moon. The Earth and moon elements were companions not only with one another but with the elements of thousands of future worlds. When their stars died and shrugged them off, they entered the next phase of a cosmos that loved to sort things out, sort nebulae into sparseness or shapes. The aluminum and titanium diverged, some joining Earth, some the moon.

The lunar aluminum and titanium lay jumbled together, with vague concentrations here and there but no lodes of pure ore. It was scrambled up with many other elements. The moon had never figured out how to separate and concentrate it. But Earth had moved beyond geological processes and achieved, with easily burnable hands, the ability to raise elements from the dead and give them purity and form. Humans admired aluminum and titanium because they were lightweight and yet strong, especially titanium, named for the Greek gods the Titans. Titanium was the best element for building airplanes and artificial joints when human bodies failed; it was welcomed into the core of human identity.

On the moon, aluminum and titanium hid their identities and strengths and accepted chaos. They had been swept aside by the lander's rocket plume, and they had sagged as a giant body of themselves settled onto them— the drive and weight of four billion years. Their formlessness upheld an elaborate sculpture of forms, its many geometries melded together, each strict shape holding its own skill and purpose, each carefully thinned wall holding apart the explosively incompatible destinations of Earth and moon.

The lander gleamed with sunlight and starlight, with elements announcing the beginning of their own long journeys.

14
GLASS

From deep inside the moon, from deep darkness, magma rebelled and rose, following faults upward toward the light. From volcanoes the magma fountained outward, far outward in the moon's low gravity, and by the time it hit the ground some of its silica had congealed into tiny globs of green glass, much of it quite round. As they flew and tumbled, the glass beads sparkled with sunlight and Earthlight. When they hit the ground they bounded and rolled and sparkled, but then they came to a rest that might last billions of years. The moon had done its best to generate form, but these eggs would not hatch any further forms. Meteor impacts fused more glass beads and splattered them across the ground and into the boulders for miles, and micrometeorites and solar flares created other kinds of glass. The moon was paved with glass. Yet these glass eyes could not see themselves or string themselves into jewelry and find themselves beautiful.

And then one long lunar day there came from the sky a revelation: two tablets of glass, flat and thin and strong, sparkling with sunlight and Earthlight and moonlight. They too were full of geological power. They too had once been silica shot or squeezed from volcanoes. They had become rock that held its shape for eons, then dissolved and flowed down rivers to the sea and piled up on the seabed and beaches and been pushed back and forth by the moon for eons, until once again they became solid rock, this time sandstone. They arose into mountains shining with moonlight, only to dissolve again and flow back to the sea again and be tossed helplessly by the moon.

And then one average Earth day a chunk of obsidian,

shiny and strong black glass forged in volcanic violence, after laying unnoticed on the ground for millennia, was picked up by hands and turned this way and that. The obsidian revealed a face looking at it and seeing itself. Eventually those hands chipped obsidian into sharper and more streamlined shapes and threw them at animals; volcanic propulsion became the violence of animals bursting apart and gushing blood and falling and being flamed. Before long, human hands turned volcanic propulsion into rockets roaring to the moon, carrying glass windows.

When humans accidentally discovered how silica turned into glass, they fell in love with glass. Glass was fragile and usually short-lived, but humans decided glass was still worthwhile, just as life itself was fragile and doomed and full of splinters yet worthwhile anyway. Glass made beautiful the light that life needed and loved. Humans made glass cups when wood or metal would have sufficed, because glass sparkled the water or wine essential for life. Women wore glass beads to bedazzle men, and kings took glass beads into the afterlife to bedazzle the gods. With stained glass windows, with halos around the god who used light to create the world, humans proclaimed the goodness of light and life and tried to magnify them into eternity. With glass telescopes, humans sought the secrets of the moon and cosmos.

With the silica in the lander computers, with moon-swept beaches become cities of electronics and calculation, the geometry of the solar system had flowed into glass and guided humans to the moon. Through windows that kept Earth inside and the moon outside and yet allowed light and vision to roam freely, through muddled earth that humans had clarified, the astronauts looked out at the moon, at glass that sparkled so faintly they could not really see it.

Nor did they notice the glass windows that held four and a half billion years of illumination.

15
THE FACE IN THE MIRROR

The astronauts opened the equipment bay on the outside of the lander and unloaded three experiments. Two of them, folded metal and glass, totaled 170 pounds, on Earth. One astronaut picked up both and walked far enough from the lander that the experiments wouldn't be disturbed when the lander took off. One experiment was a panel with one hundred small round mirrors. The astronaut unfolded it, set it up to point at Earth, and carefully adjusted its angle.

After the astronauts left the moon, a laser beam would streak out of space and find the hundred reflectors and rebound into space, back to Earth, back to the mountaintop moondome observatory that had sent it. For four billion years sunlight had shined onto Earth and been absorbed, but now sunlight had been folded a billion times into a light beam that could leap outward, through the atmosphere, and across a quarter million miles of space. For four billion years some sunlight had reflected off Earth and onto the moon, diffused and unwarm, but now Earthlight had become far more intense, charged with genius and curiosity, probing for answers. The laser beam also illuminated the ground around its target, and glowed faintly off the lander. The laser beam took a second and a half to race back to Earth, as human experience might measure it, but human instruments measured its timing far more precisely, measured it to determine the exact distance between Earth and moon. The laser would also reveal the rate at which the moon was receding from Earth, which might suggest something about their origins. Subtleties in the moon's motion might hint at the moon's internal structure.

Yet even before the laser beam hit it, the laser reflector had revealed the distance between Earth and moon, revealed it to be vastly farther than a quarter million miles. The laser reflector had revealed it with its own design, its precisely shaped metal and glass, its handle and base and adjustment lever and geometrical mirrors. The laser reflector had revealed it when the astronaut was setting up the reflector and looked into it and his shape was mirrored in it, outlined against the empty sky. The distance between Earth and moon was revealed in the difference between a living shape and formlessness. Earth and moon dwelled billions of years apart, illumination-years apart.

Earth had become a lonely quasar searching the cosmos for even a flicker of its own face.

16
LUNA MOTHS

Covered with billions of circles gouged by predators, the moon arose, tinting the forest with pale light. Across the moon, for a few seconds, fluttered a light-green luna moth. It was named for the moon-like spots on its wings, spots meant to resemble large eyes to scare off predators, owls by night and other birds by day. The luna moth's long, twin tales were meant to baffle the sonars of bats. The luna moth was using the luna to plot its path through the forest. After emerging from its cocoon, the moth had only seven days to live, seven days to find a mate and lay eggs and ensure that the luna moth form—more than four inches long and wide, four large wings, fuzzy white body—would fly far into the future. The luna moth species seemed confident of this seven-day deadline, for it gave its members no mouths with which to recharge their energy, all of which came from voracious eating as caterpillars.

The luna moth felt its energy draining away, felt the attraction of the immortal moon, the unworried moon. The moon never noticed the dramas and urges and transformations of life.

With sunset, moonflowers opened up, opened their large, deep, white flowers to the night, to the eyeless moon. Yet moths saw the flowers and came to them, not luna moths but moths with long, unrolling tongues for tapping the flowers' delicious energy. The moonflowers stared at the moon, stared like the lunar lander's white dish antenna searching for eyes that could recognize the dramas of life.

17
THEIA

The astronauts found everything unfamiliar: moving and balancing in low gravity, the impress of their boots into the lunar dust, the moon's shapes and textures and colors, the moon's response to their tools. All their training to make everything familiar, at least their equipment and activities, still left a deep, porous interface with the unknown moon. Yet the moon should not have seemed so unfamiliar; the astronauts should have looked upon it with geo-netic recognition, for the astronauts had known its rock and dust long ago, known it very well. This dust, much of it, had once been part of Earth.

Some 4.5 billion years ago Earth had become a stable planet. It had condensed out of the solar nebula and sorted itself into a heavier, metallic core and a lighter, rocky mantle, and its surface was showing patterns, if chaotic patterns. Nebular particles were still raining onto Earth, though thinning out, and asteroids and comets would continue coming in large numbers and sometimes large sizes for a long time, adding their ingredients, especially comet water, to Earth's mixture of elements, but on the whole Earth's size and components seemed well defined.

Yet some 100 million years after Earth formed, it suffered a major convulsion and change. Another planet, the size of Mars, about one-tenth the size of Earth, had formed somewhere else in the inner solar system but not in a fully stable orbit, or perhaps some gravity vector or passing body nudged it from its orbit. It wandered away, perhaps crossing Earth's orbit many times, and finally its course intersected Earth's course. It accelerated and swerved and crashed into Earth, smashing both worlds apart, shooting out debris, forming a vast, hot, swirling cloud. The matter of both worlds brewed together. In a very short time, perhaps only days, the chaos took shape: gravity and momentum sorted it into two swirls, a larger and smaller swirl, and two bodies formed, not the previous two bodies but bodies with new sizes and mixtures of elements. The metallic core of the impactor mostly sank into and merged with Earth's core, while the lighter mantles of both planets were propelled farther outward and formed the smaller world soon reliably orbiting the new Earth. The moon was a blending of Earth matter and alien matter.

For 100 million years the matter in the astronaut bodies had been thoroughly mixed together with some of this moonrock and dust, flowing with it, boiling with it, spouting with it, forming patterns with it, forming rocks with it. They should have remained together forever and gone on many adventures together, but physics and fate had intervened and sent them to separate destinies. The astronauts could have looked upon the moondust as long-lost companions and their arrival as a reunion, but they could not see this, for they did not yet know it. Only the moonrocks knew it, a secret they had kept for 4.5 billion years, and now they would share it with their forgetful friends.

If not for the impactor (named "Theia" by a species prone to disuniting the cosmos with names), matter that was now part of the moon would have gone on Earth's long evolutionary journey. It would have formed continents and wandered around, formed faults and set off earthquakes and raised mountains and volcanoes, formed many kinds of rocks and eroded into many kinds of particles and canyons, formed oceans without any tides, formed clouds and storms and rivers, formed molecules prone to disuniting from chaos and holding onto separate identities. The moondust would have become many kinds of cells uniting into many forms, become leaves worshipping the sun, become roots seeking the wayward sea, become fish and sea turtles and whales never feeling the tides, become herons and ravens and cardinals living the wind, become spiders and lizards and rabbits with the same beating heart, become tigers that felt no affinity with other lifeforms but only with their dead flesh. Moondust would have become humans awed by the night sky, become goddesses but never moon goddesses. Would moondust have become astronauts if there was no moon to ponder and reach?

Instead, this Earth matter had been exiled to the moon, where its atomic talents were not appreciated, where it was shaped mainly by external forces, violent forces, where it sat and waited to decay, sat and waited to decay further, sat and waited until one day there came from the sky Earth matter that revealed what Earth had become and what the moon had missed out on. The moon was inhabited by ghosts, all the beings who might have lived. The moon ghosts watched the astronauts moving about, their bodies so solid, their imprints so real. When an astronaut reached out to touch the moon, the moon ghosts wanted to reach out touch his solidity but could not. When the astronaut spoke, the moon ghosts wanted to acknowledge, but their voice was locked deep within rock. When the astronaut saw a promising rock, the moon ghosts wanted to hand it to him and say: *Recognize us and recognize yourself—this could have been you; this could have been the entire Earth.* The ghost moon haunted Earth's sky and, through humans, longed for the unobtainable.

The astronaut bodies contained not just Earth matter but Theia matter, matter that should have remained a part of Theia and perhaps fallen into the sun but that had been peeled off and swirled not into the moon but onto Earth. Some astronaut atoms had hovered on the narrow boundary between Earth and moon, wavering, calculating an equation of gravity and momentum and heat, and decided to join Earth, leaving their companions to become moondust. These Theia atoms had blended with Earth atoms and gone on all of Earth's adventures into order. They had traced out the vast boundary between order and chaos, between matter's inner genius and the crudeness where it usually ended up. Now the living Theia reached out to the dead Theia, Theia cells to Theia moondust.

The humans too were ghosts, vastly improbable, fundamentally impossible.

18
SIMULATIONS

They had noticed it with their first footsteps: the difference between reality and dreams. The odd thing was that they experienced the intensely real moon as a dream. They had taken their first step onto the moon many times before, in huge training halls and on outdoor moonscapes. They had watched and felt their first foot touching the dust and sinking in, noticing how far it sank in and how easy it was to lift it out and take a further step. They were noticing because this had been one of the larger uncertainties about the moon, because geologists hadn't agreed about how deep or stable the dust might be, whether an astronaut might sink dangerously deep, even in low gravity.

On the real moon, the astronauts were watching keenly for the moon to define its reality and to define the differences between reality and simulations. Instantly, the moon diverged from the simulations, not allowing a foot to sink in far at all. The astronauts noticed that the landing pads too had barely sunk into the dust, and that the descent engine hadn't dug much of a crater. Soon they would dig into the soil and find that it was much firmer than in the simulated moonscapes.

The astronauts had simulated their journey over and over, every possible aspect of it, with hundreds of hours in the most sophisticated simulation machines humans had ever built. In planetariums, they studied the stars by which they would navigate. In three command module simulators, they launched and orbited and docked and separated and ignited and flew to the moon and orbited and undocked, over and over, studying every gauge and switch and number, every program and procedure and checklist, every malfunction and every solution. In landing simulators, one of which really flew, they descended and landed and sometimes retreated and sometimes crashed. From pretend landers, they practiced climbing onto the moon in bulky moonsuits and using tools and cameras and setting up experiments. In their own imaginations they could not stop practicing. They practiced for the same reason actors practice for a play, musicians for a concert, and athletes for a game, to ingrain actions deep into memory, to make them as automatic as possible, and to make parts mesh perfectly into wholes. The simulations were directed into the tracks in the human brain that make the world wholly familiar and predictable and that leave only novelties or motions to stimulate excitement, leaving humans feeling that being alive is entirely ordinary and often boring.

Yet the real moon fractured its molds in the human mind, repeatedly. It began with a rocket launch that was far more jolting and loud than first expected. The sunlight entering the command module and the sunlight and moonlight entering the lunar lander were far brighter than in the simulators, challenging sleep, and on the lunar surface both the light and shadows were so intense they challenged focus. The moon's color was so unique that the astronauts couldn't agree on how to describe it. As the lander neared the surface, the dust streaked outward in a way no one had anticipated, confusing the astronauts' sense of motion. The stars projected in the lander simulator windows, stars essential for liftoff navigation, were not visible from the real moon. The astronauts needed far more time to prepare for their moonwalk than planned. Walking in moon gravity quickly proved easier than feared, but bending moon-glove fingers and using tools proved harder. The lunar surface was rougher than expected, making it harder to set up instruments with the necessary balance. With textures and colors, with light and shadows, the moon diverted humans into its own reality. The astronauts noticed the differences between reality and simulation, and sometimes commented on them

In truth, humans spend a great deal of their lives immersed in simulations. Their nights are made of dreams, and much of their days are lost in dreams, identity fantasies of being somewhere else or someone else, of having more status, love, money, looks, or talent. Humans immerse themselves in the simulations of storytelling, books, plays, movies, and television. Nations define themselves as stories, even if stories contradicted by the stories of other nations, contradicted to the point of war or of flying to the moon. Humans gaze into the sky and into their own struggles and deaths and imagine thousands of gods and plans and meanings, and they arrange all their actions and perceptions and feelings and their grandest architecture to fit into those cosmic simulations.

Yet to the astronauts, the moon was pointing out the difference between human dreams and reality. The moon was so far outside of human experiences that human brains could not match it with previous patterns, could not re-cognize it, and thus the astronauts perceived the moon as unreal, surrealistic, a dream. They were being tricked by brains designed to color the world as familiar and unexciting, brains programmed to notice the new but not the normal, brains programmed for survival but not for living.

Trying to sleep inside the lander, the astronauts were kept awake by cycling valves and pumps and fans they had not bothered to simulate because no one had thought it important. Even here, the moon was trying to awaken humans from their trance.

The moon was saying to the moon strangers: there is so much reality you have never perceived. The moon showed them a patch of ground with pebbles and a collage of vivid light and shadow and asked them if they had ever really noticed the ground and the light and the shadows. It showed them their own shadows and suggested that something must be casting those shadows, making them move. The moon announced: this is not a test. This is yourself, this is the universe, intensely real and wondrous and strange. You were surrealistic all along and now you have glimpsed yourself truly. Look.

19
GHOSTS

As the first astronaut stepped down the ladder to the moon and took his first steps upon it, as both astronauts moved around the surface, they were shown on television screens all over Earth, yet shown strangely, at first upside down, then vague and streaked, too bright or too dark. When they moved, they blurred. It seemed you could see right through them. They looked like ghosts.

There were, of course, technical explanations. The television camera, which had to work in a vacuum and in unfiltered sunlight, was cutting-edge technology, but it did not scan images nearly as fast as television sets showed images, so moon images had to be translated by repeating them five additional times. When the astronauts moved too fast, they left their repeating images behind, becoming blurs. The camera couldn't do justice to both the deep shadow behind the lander and the glare on the sunlit ground, and the first tracking station to receive images was fumbling them, turning everything into deep shadows.

Yet it was appropriate that the moon images had an unearthly glow to them, that the moon seemed a twilight zone, and that the astronauts looked like ghosts, for they truly were ghosts, not the identities they had imagined to be so solid on Earth but ghosts all along, brief lives flickering through a vast cosmos, impossible beings walking upon a dead world, strange lights and motions in a cosmos that should have remained dark and still, apparitions come from the sky, haunted by unresolvable questions.

20
THE FORM

Seen from Earth, the moon was without form, and vague. Clouds rolled across the moon so thickly that it wasn't even a circle in the sky, just a blurred glow in the clouds, fluctuating in shape and brightness. The clouds were carrying the oceans to the mountains, carrying rain to the forests and fields, carrying life into the future. The moon pulsed like a jellyfish, like a white moth, like a white heart. The clouds thickened and thinned, finally thinning enough to let the moon appear as a circle, a full moon. The moon revealed silhouettes before it, for it was a rising moon, just above the horizon but not above a tree on a hillside. As clouds streaked across the moon, the tree flickered in and out of sight, flickered in and out of the clouds as if it was condensing from them—as indeed it had. The moon became more visible, and another shape streaked across it, wings flapping, an owl with eyes intense with moonlight. Then another silhouette appeared and moved up the hill and stood beside the tree, a body with two legs, two arms, and a prominent head. This shape flickered in and out of further clouds, as if it was condensing from them—which indeed it had, just as it had condensed out of interstellar clouds, just as it and the moon had condensed out of the streaking energies of the Big Bang. The human stood quietly, a dark shape in the night. The moon recognized none of the social identities the human wore by daylight and considered so important; the moon revealed only the human's elementary shape, its truest shape, the shape of a cosmos flowing into ever stranger energies and orbits and shapes.

21
CARBON

A carbon boot stood upon a carbon desert.

The element of life was very scarce on the moon. Most of what was there had arrived on asteroids and comets long ago, and the heat of impact had turned it into graphite. The solar system had written its history in pencil on the moon, written it in code for a locksmith to come along and lubricate open. Tiny amounts of carbon still arrived in the solar wind, but most of this soon evaporated into a lunar atmosphere so tenuous that one human could breathe in and out the entire atmosphere in a few years.

Yet if the solar wind carbon fell onto the ancient graphite carbon, it might stick and stick around. Carbon is the most gregarious of atoms. Most atoms have limited dockage and can form only small molecules. Carbon loves to bond with other atoms, especially itself, and can form millions of molecules, some with millions of atoms. Carbon was the only element with the complexity and agility necessary for life. Carbon's obsession with making order was at work even on the moon, as lunar carbon recognized its brethren arriving from the sun and tried to grasp it.

Now the lunar carbon was having a revelation. A solid mass of carbon was standing upon it. The astronaut's boots, suit, gloves, helmet, air hoses, and control knobs were made of varied forms of carbon, all united to protect the carbon palace within. As the carbon moved, imprinting its shapes onto the dust, the carbon in the dust reached out to the carbon in the boot, trying to grasp it, and perhaps an atom or two touched fingers, but otherwise the chasm was too enormous to cross. The lunar carbon could not join the league of Earth carbon.

From all the carbon the astronaut was wearing, nylon and rubber and plastics and composites, carbon was outgassing. The air pressure inside and the vacuum and heat outside encouraged looser atoms to detach themselves and drift away. From the lander, too, carbon was outgassing, following the burst of carbon dioxide that escaped when the vents and door were opened. The carbon atoms, feeling no loyalty to the moon, wandered upward. As they rose, a few atoms encountered lunar carbon atoms and they all forgot their past identities and behaved simply as carbon, dancing around one another and joining electron hands and forming a new molecule, an assertion of order even as the Earth carbon was evicted from ultimate order and exiled back into the formlessness and loneliness of space.

This outgassing did not include the carbon dioxide the astronaut was breathing out, which was not vented into space but scrubbed by filters, the mechanical equivalent of green plants, so that his life support backpack's limited oxygen supply could be re-breathed and maximized. His breathed-out carbon remained gummed within the backpack, and thus when the backpacks were tossed overboard to lighten liftoff weight, the carbon that had energized the moonwalk remained upon the moon. Slowly, as the backpacks aged, as the extreme heat and cold stressed them, as their plastic tubes and rubber seals wore out, the astronaut breath would begin outgassing, being breathed out again. Their carbon atoms, some of which had been part of them for a long time, formed a vague shape right where they had stood, wavered right where they had walked, wavered like a ghost.

22
MOONQUAKES

Suddenly, sunlight was reaching all the way to the center of the moon, 1,080 miles through deep, solid, dark rock.

For four billion years, sunlight had touched only the moon's surface, though its heat may have reached a bit deeper, fracturing rocks and shifting dust. The sunlight had been a crude and blind force. Yet now the sunlight was landing on a glass and metal leaf that organized and channeled it and let it peer into the moon, let it feel vibrations from throughout the moon, including rock falls triggered by its own heat.

The astronauts had set up a seismometer to probe the moon's deepest secrets. Was the moon like Earth, with a molten core and tectonic plates and earthquakes? On Earth, earthquakes could x-ray Earth's inner structure; what would moonquakes reveal about the moon? Were moonquakes triggered by inner forces, by gravitational stress from Earth, or by meteor impacts?

As soon as the seismometer was turned on, it began registering moonquakes, small but very steady moonquakes: quake-quake, quake-quake, quake-quake. This was indeed a deep secret, an ancient force. The moon had never felt a steady rhythm like this before. The seismometer was registering a force that had arisen out of a magma core, out of tectonic plates and earthquakes and lava flows, out of meteor storms and thunderstorms and wonder storms, a force now walking upon the moon.

Two days later the seismometer registered a meteor impact that sent waves of energy throughout the moon, back and forth, diminishing only slowly. This meteor gouged out a crater like all the others, another dot on the graph of mass and velocity. Yet this meteor was making its second landing on the moon. The lander had lifted off and, its mission complete, been released to fall back onto the moon, to ring the moon like a bell in the silent and oblivious night. The lander's remaining water, the ancient ocean, splashed onto the lunar beach. Its remaining air gasped in surprise.

23
MOTION

Stillness. Stillness. Stillness. Stillness. No motion at all. Complete stillness, including complete silence. Tranquility. A sea of tranquility without any waves, or only grey waves that took 100 million years to roll. Once, the moon had been far more active, from within and without, from lava welling up and asteroids crashing down, but this activity had largely died out, and now any one place might wait eons for another asteroid to arrive. For most of the moon for most of the time, the only motion was the very slow decay of rocks into dust. The moon was a Zen garden for the universe to realize the concept of stillness.

Yet one eon, motion appeared in the sky, not the steady motion of planets but a far more complicated motion, a light that grew bigger and brighter and more convoluted, motion that speeded and slowed and descended, skimming over the stillness, hovering, lowering, finally joining the stillness of the moon.

Motion became a door opening, and shapes emerging and descending. Motion moved upon the stillness, nudging the dust and rocks into motion.

Motion had arrived from a world of motion where motion was normal and multiple and vigorous, where motion had flowed nonstop for more than four billion years, folding itself into ever new forms of motion, now into the motion by which a world of motion was meeting a world of stillness. Motion looked at stillness and found it strange and forbidding: stillness meant death. Motion was not converted to the moon's religion but continued being itself.

The humans were able to move because they were unleashing the motions within them, expressing layer within layer of motion: electrons whirling, molecules combining and splitting, cells growing and birthing, organs performing many activities, blood flowing, light flowing through eyes, sounds flowing through ears, and air flowing through noses and lungs. In brains, motions became awareness and ideas and emotions. All these motions became the motions of legs and arms and heads, the motions of bodies across the ground.

Their footsteps contained the motions of all of Earth, billions of years of motions. They contained the swayings of primates through the trees, the gallopings of animal herds, the global migrations of birds, the oscillations of fish. They contained the first cell that figured out how to combine motions in just the right way to keep them moving forever. The astronauts contained oceans and storms and volcanoes. The astronauts' motion was a further elaboration of the motion of Earth through space. It was propelled by the reliable cycling of the planets and the more random flights of asteroids and comets. It contained swirling nebulae, the swarmings inside stars, the explosions and collapses of stars, and the motions of stars around the galaxy. Their footsteps were a continuation of the Big Bang itself, another of the many masks it had enfolded itself into.

Throughout the universe the motions unleashed by the Big Bang continued vigorously, yet in a few pockets they had slowed down to almost nothing, to planets and moons of barely changing stone. On Earth, motion had journeyed into another extreme. Now the two extremes were intersecting: Earth was stirring the moondust. Humans embodied so much motion, defining motion as necessary and good, that they could not feel much affinity for the moon. They knew that someday, in spite of their best motions, their motions would cease and their bones, designed for motion, would lie still for centuries, as still and grey as the moon.

The astronauts walked. The cosmos that could have been stillborn, that could have contained nothing but stillness forever, was glorying in contemplating the stillness and stirring the dust.

24
CHAOS

As the lunar lander was coming down to land, it was heading straight for a deep crater hundreds of feet across, surrounded by boulders and long streaks of rocks. The lander was designed to be a mediator between order and chaos, between Earth and moon, to fit life into the roughness of the moon. It had legs that were tall enough, spread enough, and compressible enough to land on uneven ground, and had landing pads wide enough to support its weight upon deep dust. Yet now the lander was heading for more chaos than it could handle, steepness and roughness that might break it or break it apart. The lander's four legs represented the legs of the humans inside, who on Earth operated by standing upright, keenly balanced, with no tilt except the carefully controlled tilt of walking, and even here on the moon, with less gravity to tilt them, humans needed to replicate the order of Earth. The lander was heading for the crater because it was on automatic pilot, a brilliant, order-packed invention for serving order that was now serving chaos because it was working blindly. Order-packed human eyes swarmed with light; order-packed brains swarmed with images and recognized excessive chaos and began searching for more compatible patterns; order-packed hands turned off the automatic pilot and gripped a rod and pushed and pulled and tilted it, little motions that got translated into the motion of the lander, little brain firings that got translated into rocket firings. Yet the astronauts could not recognize their planned landing area, for they had already overshot it, left it a few miles behind. Even surrounded by the greatest order humans could devise, by hundreds of gauges and switches and lights and levers and a computer for measuring chaos and enforcing order, chaos was breaking out, chaos was breaking in from the chaos outside. Chaos had erupted when the lander had first separated from the command module and been unable to communicate with Earth, until a switch was moved and threw chaos back. Chaos was erupting now as an alarm light and buzzer went off, a major alarm, one the astronauts did not recognize, what it meant or what to do about it, and almost none of the experts on Earth recognized it either. Chaos was sneaky, searching for the most obscure openings. The astronauts overrode the alarm, yet it soon flared again, still a puzzle. Chaos also flared within the astronauts. They searched for a smoother landing spot, acknowledging that life was a special and fragile thing in a universe of chaos. As they searched, their fuel was running out—forty seconds, thirty seconds. The rocket began stirring dust off the ground, which meant they were getting close, but it also meant they could no longer see the surface, and the dust was streaming sideways in a way it would not in Earth's air, a streaming that made it hard for the astronauts to discern the difference between the dust's motion and their own motion, confusing them into drifting sideways when they did not intend to, drifting toward scraping across the ground and breaking the lander's legs. Time was running out; chaos was pounding on the door.

Over the ancient ruins of the moon passed the contest between order and chaos, revealing to the moon that it was not the only reality, that elsewhere there had arisen a force that had devoted itself to evading chaos, to inventing ever better forms and strategies for defeating chaos, yet which even now, after four billion years of work, even now in its most sophisticated form, was still swaying precariously, barely above defeat.

It was an ancient story, the fundamental story of

the universe, the story that began with the Big Bang and changed forms and characters and details many times but remained the same story of order contesting with chaos. From its first moment, and perhaps before its first moment, the universe was rich with impulses toward order, with particles that fit together and forces that inclined them to fit together. At birth the universe was pregnant with many thousands of kinds of forms, not just possible forms but inevitable forms. Yet the universe was also full of chaos, forces that interfered with the ordering forces, blocking and limiting and distorting them, allowing order to emerge only here and there, only partially, with chaos always ready to knock order back down. The universe's ordering forces had to play on a field swarming with randomness. It might be inevitable that elementary particles would combine into atoms, but there was no destiny in which specific particles combined with one another. The universe might be destined to create galaxies, but there was no predicting which atoms would join which galaxies, or the numbers, sizes, shapes, and contents of galaxies. Galaxies might be destined to generate stars, but they did so through a vast tug of war between billions of centers of gravity and momentum, generating stars with vastly different sizes, functions, and fates. The forces that tried to generate planets were sorted into diverse outcomes, with small factors making big differences in a planet's or moon's size, composition, temperature, and activity. Some planet formation would fail completely, and some planets and moons would be erased by star collisions, supernovae, or black holes. The universe had to generate vast numbers of planets to allow a fraction of them to have the right ingredients to move into the next adventures into order. The blueprint of life had flickered within the Big Bang but billions of years were needed for the story of order and chaos to develop from atoms to oceans and then dress itself in cells. Building life was even more tenuous, often resulting in lifeless planets or limited success or catastrophic reversals, and it was subject to a vaster lottery of circumstances, making no species inevitable, leaving a constantly changing assortment of species.

Every new life was a new test of life's ability to unfold order and outmaneuver chaos, to organize a body into the correct shape, with correctly functioning heart and nervous system, and to give it birth. Every life had to fend off chaos from within and without, had to unfold its genetic plan and to smother diseases and avoid accidents and natural disasters and lack of food and societal chaos and many forms of violence. As if life didn't face enough chaos, it inflicted a great deal more upon itself, with animals attacking one another and tearing one another apart. This too was a manifestation of cosmic chaos, for the universe had implanted within living bodies its laws of energy and dissipation, turning physics into hunger, forcing animals to constantly find new sources of energy, to grow teeth and claws and fierceness, to turn stellar fires and DNA weaving into murder. The same laws that made stars explode also made tigers leap onto antelope. The universe's ordering impulses also turned into sexual impulses; the "desire" of nebulae to form nuggets now gravitated living bodies toward one another. The universe's quest for order also inspired humans to search for order in their experiences, to invent the god Apollo to rule the sun and assure reliability to the cosmos and yet acknowledge its undeniable chaos. Humans also searched for order by gazing at the moon and imagining all sorts of meanings for it, and by reordering the forces of nature into a system that could fly to the moon so that humans could ask the moon in person what it meant.

It meant chaos. Even in the most sophisticated machine humans had ever built, chaos was invading. Humans had trespassed deep into the realm of chaos. Their instrument panel was full of alarm lights to warn about the whereabouts and activities of chaos, red alarms because that was the color of blood and fire, which could be powerful servants of life but also imminent dangers to life. The fire pouring from the lunar lander was Promethean fire, stolen from powers far greater than humans. The alarm lights essentially read: "Chaos erupting," "Cosmic chaos invading," "Monster attacking sun god Apollo." The instrument panel also held many buttons and switches that essentially said: "Repel chaos." The astronauts were reacting to chaos in wholly new ways, yet also very ancient ways. Even as they had learned from bad experiences and redesigned systems to defend order, chaos came up with new ways of attacking. Subsequent moonflights were hit by different crises, nearly ending lives.

In the end, of course, chaos would win. Even if the astronauts completed their missions and came home alive, chaos would stalk them, casting shadows that never left them, even in the night, eroding their bodies and minds, inflicting them with disability and disease, finally pulling them out of life and back into chaos. Their granite tombstones too would erode. The footprints they had left on the moon would blur and disappear. In the end, chaos would burn Earth into ashes, burn out the sun, and leave the entire universe dark and cold.

Yet in the meantime, life would enjoy being itself, make the most of what it had been given, enjoy the feel of sunshine upon its face and the look of the stars and the touch of the richly textured earth. Life would paint murals in caves and perform music and build cathedrals and fly to the moon to challenge chaos, to tease it for its limitations, to define life's own abundant realm.

The universe's long journey from chaos into order had finally become the journey of a spacecraft to the moon, the journey of life to encounter the chaos from which it had arisen and which it still contained. The flash of the Big Bang, the ignition of stars, and the glow of galaxies had turned into the careful fires of a Saturn 5 rocket and the lunar lander engine. The emergence of order, the emergence of galaxies from nebulae and of trees from seeds, had become the emergence of a piece of Earth into space. The lunar lander embodied 13.8 billion years of work, the welding of the stars and the bolting of tectonic plates. The voices within were the voices of volcanoes and surf translated into a more articulate language. The lander's balancing maneuvers as it approached the moon were but the refinement of a balancing act the universe had been performing all along. The lights on its instrument panels were what the stars had focused themselves into. The astronauts' fingers were comets that had slipped into gloves and were now pointing more knowingly. The astronauts looked through the newest stained glass windows for celebrating a universe aglow with light. The oblivious energies of stars and rocks looked through human eyes and felt recognition and joy. The moon landing had completed a circle, performed a confirmation that the journey from the Big Bang had been good.

25
CONNECTIONS

Surrounded by airlessness and silence, encased in a goldfish bowl of air, in solitary confinement, the astronaut spoke as if nothing were unusual, and the air, well-trained back on Earth, the air that had been carrying voices for hundreds of millions of years and before that carrying the voice of the tides and thus the moon, the air carried his words only a few inches until it hit a dead end. Yet the ripples of air were translated into ripples of electricity, which could cross a vacuum, cross to the other astronaut a few feet away, where they were translated back into ripples of air, ripples of hearing, ripples of consciousness. His voice was also sent to the lander, from which it was sent to the command module fifty miles above and to Earth five thousand times farther away. Landing in a giant dish antenna, the voice was shot back into space to a satellite, then back to Earth and across an ocean to another dish, then through cables to ventriloquist "speakers." The astronaut's voice joined a world flooded with voices, the voices of millions of species, incessantly communicating their presence, their awakenings, their feelings, their discoveries of food and danger, their maps and weather reports, their core need to make connections with other lives.

The astronaut spoke into a bubble of air severely disconnected from its mother lode of air by a quarter million miles of emptiness and silence, which was only an inch of the emptiness and silence that filled the universe and that had been growing ever deeper since the Big Bang. The moon belonged to that emptiness and silence and was far more normal than was Earth. The moon rocks felt no need to tell each other anything. The moon rocks were content to be connected only physically. Now among the rocks had arrived a creature venting its manic metabolism partly through speech, partly by searching for connections. Accustomed to connecting with a world of humans and machines, the astronauts looked at the faceless, leverless moonscape and had trouble relating to it. To the silent moonrocks they said nothing. They met the moon by applying familiar little machines to it. They steadily maintained their word lifeline to the jabbering planet. Still, their radar minds were scanning, recognizing things familiar from Earth and ideas become real, applying the names that made things live, that made things safe, making connections, weaving the slender thread of Earth voices onto a moon no longer silent, no longer adrift.

26
THE MOONLIGHT SONATA

Odd: the silence of it. The astronaut reached out with the tool and scraped it across the ground, pushing it harder to dig deeper. On Earth he would have heard the scraping of metal against rock, heard a confirmation of his own effort, but here there was only silence. He heard his own breathing, perhaps a bit stronger from his effort, but heard nothing from the ground, from Earth metal touching moon rock to find out what it was made of. The moon was made of silence.

He reached out his hands and spread his fingers upon the elephant tusks, which once had dug into the earth and found it nutritious. He pressed down on the keys, the keys to a door that on the moon remained forever locked. Through the keys he was activating wood that once had swayed and rustled in the wind, and metal that once had rumbled as earthquakes. With hands made of ocean surf and thunderstorms and bird voices he set the air vibrating, sending wave after wave spreading across the room, waves alternating in length and strength. A room full of molecules jumped up and down and back and forth to conduct the waves onward. Each molecule didn't travel far, only very locally, only enough to hit its neighboring molecules and set them jumping with the same energy. It was only because the room was too crowded with molecules that the waves rolled onward, only the waves, only the energy, only the patterns. When the waves met the hard waves of the human ear and set its filaments rippling, the mere jostling of molecules was translated into consciousness, into music, into molecules jostling with joy.

Other planets and some moons know all about atmospheres and about carrying sound waves. Even the sun holds sound waves. Even the Big Bang, with its super density, carried special acoustic waves at sixty percent the speed of light, waves that made a major imprint on the distribution of matter, leaving today's galaxies concentrated in walls of galaxies. The music of the spheres, indeed. Yet the moon never got wind of this concept; the moon remained locked in silence. Even when the astronauts were standing inches apart, they had to translate their sounds into silent waves of electricity and decode them back into sounds. Their helmets were tiny bubbles of sound amid the vast silence of the moon, bubbles of air that for billions of years had carried the sounds of wind and dinosaurs and birds. The astronauts' breathing continued very ancient sounds. Their voices were made of water that had fallen over cliffs, land that had shouted as volcanoes, and air that had heard the moon's tidal energies and given it voice. An astronaut spoke the word "moon," a symbol for a world without symbols, and the word flew upward and across a quarter million miles of silence to where it became sound and meaning again. Yet the moon, even with its billions of ear-like craters, could not hear its own name.

He moved his hands over the piano keys, slowly, gently, turning cellular vibrations into air vibrations, turning the air that gives life to humans into a new kind of life. Humans love music because it is made of patterns and so are we: music plays with our cell-deep sense of order, building patterns, repeating reliable patterns. The music flowed across the room, only the waves, only the patterns, just as human bodies soon discard their atoms and recruit new atoms to carry onward the same pattern for a lifetime, just as the patterns discard their bodies but create new ones to carry themselves through the generations and the eons. From the deepest wellspring of order, the music flowed across the room. It was gentle music, spacious, subtle, pensive, gently rippling music. People who heard it would agree with its title and imagine moonlight rippling on a lake, moonlight dreaming onto a mountain, moonlight encouraging lovers to perpetuate patterns, moonlight making the harsh world more gentle and mysterious.

The music flowed, but the pianist could not hear it, for he had gone deaf. He would never again be able to hear his own compositions. But he continued composing, for the pattern-generating force within him was too powerful to shut off. It had been generating patterns for billions of years. The silent and deaf moon had reached out across the mother silence from which even he had been born and seized him and used him to articulate its own deeply buried yet deeply powerful music.

27
IRON

The iron-rich rocks had sat there for billions of years, rough shapes with rough textures, without symmetries or smoothness. Were these rocks part of the moon's original crust, excavated by meteorites? Or had they been part of the meteorites, smashed apart and scattered afar? Scattered roughly, barely justifying the word "pattern," only more rocks here and fewer there, larger rocks here and smaller there. The iron rocks (about one-eighth of the moon was iron) never rusted or turned red, for there was no water or air to add further corrosion to the corrosions of space. The iron rocks sat there, strong yet victimized.

And then one day another form of iron was set down amid the rocks. It was iron blended with chromium into steel, stainless steel, for it came from a world where iron had to worry about staining and corroding, where resistance to corrosion had generated numerous gods and temples. There, too, iron meteorites fell from the sky, but they became not just holes but holy, gifts and signs from the gods, the center of pilgrimages and rituals. This steel had four smooth sides, rounded at the corners, giving the confused moon four sacred directions. It was shiny, but a mirror in which the moon could not see itself. The steel had a lid, which now opened, releasing a bit of oxygen that had done its best to corrupt living bodies but had been used by them to fight corruption. The lid revealed an empty space within.

The moon rocks were lifted by another form of iron, iron flowing through veins, iron flowing in muscles and brains and eyes, iron seeing the rough iron of the moon, iron lifting asteroids from their stillness and giving them a new trajectory, god iron lifting moon rocks from their graves and giving them a new incarnation. The moon rocks were packed geometrically, and the lid with its hinges and rods and seals was lowered and sealed, along with a bit of lunar vacuum, soon to be released on Earth.

28
POTATOES

Earth's crust rifted apart, stretching and thinning along a sixty-mile line, and cracked open. Magma poured out, forming volcanoes and wide deltas of lava, forming clouds of ash that settled many miles away and built plains of ashy soil. Over the millennia the lava decayed, adding to the soil. The land was dark and desolate. Humans called it "Craters of the Moon." At night, the craters glowed faintly from their sister craters in the sky.

One day came some astronauts. They walked over the bulging, cracked lava flows, crunched over volcanic cinders, climbed into a lava cave, climbed a volcano and looked into its crater. They were seeking the secrets of the moon, the inside turned outside, the ancient still present. They examined several types of lava. They picked up rocks and turned them over and tossed them back onto the ground. Even on the barest areas, they noticed green lichen. They saw pine trees growing out of the lava. Miles away, they saw a boundary where the craters and lava flows yielded to green fields.

The green was feeding sunlight to the potatoes under the surface. Volcanic soils were the best for growing potatoes, for its ashes were rich with minerals, and its looseness sponged water and left room for tubers to expand easily. The climate around Idaho's Craters of the Moon National Monument, with warm days and cool nights, left potatoes big and solid and dry. Minerals that on the moon lay void and abandoned were turned into roundish, nutritionally rich, white moons. Thus it was plausible that a few of these potatoes, after months of growing anonymously alongside millions of others, might be inspected and selected for a special mission. They were cleaned and cut and cooked and dried and slipped into plastic pouches with other elite vegetables. They were boarded onto a jet, flown to the coast, loaded into special boxes, elevated hundreds of feet, and, with violent stirrings, shot into the sky and into space. They headed for the moon.

Circling the moon, some of the potatoes were mixed with water and squeezed into mouths that tested them for safety and value and found them good, squeezed into stomachs that dissolved them and distributed them throughout bodies and into eyes and brains from which the potatoes looked upon the moon that had watched them growing. Some of the potatoes landed on the moon and were eaten there, and they helped energize the astronauts in putting on their moonsuits, climbing down the ladder, and taking one small step onto the dust, which the potatoes might have described as one giant leap for potatoes but which the humans overruled, for the humans had instantly forgotten their identity as potatoes, just as they ignored their identity as everything else they contained— as soil and sea and volcanoes, as eons of life and stars. The potatoes looked at the craters of the moon. The potatoes energized wonder at the moon and at moving with so little weight, and energized worry about injury and time and equipment. The potatoes looked up and saw the potato in the sky.

Carrots that could have been eaten by rabbits months ago walked back and forth on the moon. Like a sun, a juicy orange glowed upon the moon. A salmon swam upon the moon, swam far upstream. A cow jumped over the moon.

Out of its own craters and lava and void, out of its dissolute minerals, out of all improbability, Earth had risen into a feast of identities, risen into a glance back at Earth, at a place where even the most barren and black ground held patches of green lichen.

29
AURORAS

The astronaut walked to a spot close to the lander but well away from its shadow, a spot bright with sunlight. He unfurled a banner of metal foil. He would leave it there for about an hour, then roll it up and take it back to Earth. The banner was collecting particles streaming from the sun at hundreds of miles per second, particles that would reveal the sun's composition and activity. It was this "solar wind" that generates the tails of comets, always pointing away from the sun. The solar wind couldn't be collected on Earth because Earth's magnetic shield deflected it, but the moon had nothing to stop it. The astronaut stood there against the clear black sky, in the breath of the dragon.

On Earth, the solar wind becomes alive. Earth's magnetic field funnels it towards the poles and swirls it and collides it with the molecules of the atmosphere, generating light, different colors from different molecules, green and blue and purple and red, lights that pulse and dance across the sky. The green is the footprints of the oxygen that also makes animals pulse and dance. In auroras, humans see the magic and mystery of the universe. Australian aborigines see auroras as the campfires of the spirit world. Inuits see the spirits of the dead dancing happily; Inuit shamans journey to the moon to better interact with the spirits and perhaps receive help in healing. The solar wind blew into human nostrils and drove the weather inside human minds into awe.

On the moon, the solar wind had only dug tiny, dusty graves for itself—until now.

30
MAGNESIUM

The astronauts walked on soil rich with magnesium and made impressions in it, but moon magnesium did not recognize the magnesium inside the astronauts. The magnesium of moon and of humans was chemically the same, and it had gone on the same astronomical and geological journeys. Yet the magnesium of Earth had been enlisted in a further cause and served it loyally. Cells recruited magnesium to make their energy flow. The earliest plants, such as blue-green algae, packed magnesium into chlorophyll to capture sunlight. In animals, magnesium served in DNA, nerves, and muscles. Astronaut magnesium was digging up moon magnesium.

Not many years before, magnesium had proven its loyalty to life by packing itself into metal berries and diving from the sky. Magnesium was biologically valuable because it was volatile. When the magnesium hit the ground, which was obstructed by homes, shops, schools, and churches, the magnesium exploded and sprayed outward and burned intensely, unstoppable by water, destroying in one fire-constellated night thousands of buildings and tens of thousands of bodies in which magnesium struggled heroically to serve life loyally, until the very end.

Moon magnesium did not understand Earth magnesium, its sense of loyalty. Moon magnesium was loyal only to itself and entirely indifferent to life.

31
THE OWL

Who?, asked the owl. *Who are you?*

The owl was looking across the lake, speckled with stars, at the line of pine trees, at the glow behind them. The glow brightened and became a sliver of light, then a dome, then an egg, outlining the trees, then rising above them. The owl watched the moon and asked: *Who are you?*

The owl could discern moon features far better than most animals. She saw hundreds of craters in sharp detail, their edges, rays, textures, and colors. She welcomed the moon, and not just for the light it cast onto her hunting grounds. She welcomed the moon into two craters that loved light for its own sake.

The moon glowed on the water and grew a tail of light, a tale of light.

The moon looked at the lake, at the trees, at the owl, looked through millions of eyes large and small, deep and shallow, looked with all the elements that had formed eyes on Earth, but the moon saw nothing. The moon's eyes were empty.

The owl watched the moon rising, watched the changing light and shadow on the water and in the woods. The owl watched a formation of geese flying in front of the moon, signing the moon with the forms and energies of Earth. The owl watched an airplane flying in front of the moon. The owl watched a human walking along the distant lakeshore. Humans belonged to the sun; owls belonged to the moon. The owl saw the human face revealed with moonlight, disguised with moonlight. The owl asked: *Who are you?* The owl was hunting for God.

32
A VISION

The Andromeda Galaxy said *Let there be light,* and there was light. And the light flew into deep space, away from Andromeda, leaving it floating there like an angel looking for lives that needed light. The light flew onward for a million years, two million years, and would have flown onward forever except that another galaxy drifted into its path, and once again the light was flying among stars. The light ran into a planet, dove into its atmosphere, fell towards the shoreline between a coral reef and a jungle, and fell into the eyes of an amphibian not yet asleep. The Andromeda light triggered brain energies and made imprints. After awhile the full moon arose and flowed into the amphibian's brain.

When the amphibian died a few months later it settled onto the seabed and was buried under thick blankets of algae and plankton and other organic debris, and it remained there for more than 200 million years. The light of Andromeda and the moon turned black. Then one day the amphibian was summoned back to the light and transformed into a long, thin, rolled strip of carbon. It was carried to the moon.

The moon craters recorded the sunlight only by warming up, by vibrating their dust molecules at an increasing rate, then cooling down again. They retained no memory of another day, no memory of millions of days. But one day a new crater appeared on the moon, a crater perfectly round and made of metal. The sunlight fell into the crater and, with a click that was totally silent, the crater opened for an instant and let some light into a cave within, where it energized Permian green leaves and made imprints in it. The moon craters became images of themselves. The metal crater opened again and this time a speck of the eye of an ancient amphibian helped see a white thing standing amid the craters, a whiteness with its own light-hungry, shape-loving craters, a whiteness in which the light of the moon and stars and galaxies and Big Bang had been transformed into a giant question and into the answer "And God said *Let there be light*, and there was light," a whiteness standing there like an angel offering affirmation, yet a whiteness still questioning the darkness and still searching.

33
SALT

They were the salt of the earth, walking upon the salt of the moon.

Melded into the moon rock and dust were sodium, one of the moon's major components, and small amounts of potassium, once locked deep inside the moon but sprayed out by volcanoes. The sodium and potassium were mingled chaotically, sitting where they had sat for eons, achieving little.

Above them, sodium and potassium were performing elaborate dances and duties. Billions of cell membranes held pumps that regulated the balance of sodium and potassium between cell insides and outsides. Through special gates, special molecules carried three sodium ions out of the cell and two potassium ions into the cell, maintaining an electrical gradient that allowed cells to function, muscles to contract, hearts to beat, and neurons to fire. Three out and two in: the cells counted precisely and rearranged continuously. The cells ran on autopilot, their own domains, with no conscious directions from the brain. Larger and more elaborate systems regulated the balance and flow and application of sodium and potassium throughout the body, including the chemical formula of blood, the pressure of blood flow, the flow and potency of hormones, and the performance of organs. The astronauts' vision of moon and Earth was generated by sodium and potassium pixels. Their awe was a sodium-potassium drug. To assure that oblivious humans continued ingesting enough sodium and potassium, the taste buds gave pleasure signals: *This is good! Give me more!* The sodium and potassium were one of numerous layers of order that supported life with extraordinary precision and skill, and life was one layer of order amid atomic, geological, and cosmic orders that were also extraordinarily structured. Even before science, before humans saw all the details of cosmic order, they saw the improbability and the gift of life and felt there must be a giver.

On a later moon landing, one astronaut, removed from all human things and strivings, naked before the stars and Earth and ancient lunar mountains, felt the reality and presence of God as he had never felt it on Earth. Did he also feel his heart going into a seriously irregular beat due to potassium depletion? He was working hard, a nearly seven-hour day of collecting rock samples, getting hot and sore and tired, sweating potassium. In his training right before the launch he had been working hard in the summer heat and gradually draining potassium, and now, with his water sipper broken, he became severely dehydrated and his potassium level dropped below where it could support a normal heartbeat. He was overheating now because the ancient heat of lunar lava flows had been resurrected within him as he worked upon it harder than anticipated. He could have had a heart attack right then on the moon. He could have fallen onto the potassium that knew and cared nothing about sustaining life. Earth and moon were not so far apart after all, for Earth still contained the moon; life still contained the dumb chaos it had been trying to escape for billions of years but could evade only fleetingly before falling back down, onto the moon. For this astronaut, this was the beginning of years of heart troubles. His first heart attack came two years after his moonwalk, and his third heart attack killed him. He had spent his post-astronaut years testifying about God's presence on the moon, about how the cosmos was a contest between goodness and evil, order and chaos, falling and resurrection.

34
ZEROS

The moon was counting. Zero. Zero. Zero. Zero. The moon was covered with zeros of every size. In the cosmic game of balls swarming about and colliding with one another, the moon was a giant scoreboard and the score was zero, zero, zero, zero. No one was winning. The game had gone into overtime four billion years ago. The players continued destroying themselves. No one had scored any points. The game was pointless. The game left countless bodies, formed through great time and material and skill and energy, smashed into boulders and molecules. It left countless larger bodies saturated with craters yet never satisfied with enough craters—always new craters were cutting and obliterating old craters. It left the universe swarming with chaotic, dead, blind bodies pointlessly reshuffling their dust back and forth, back and forth, up and down, back and forth, forever. When the moon added up all its zeros, it added up to zero. When the moon multiplied all its zeros, it still amounted to zero. When the moon divided all its zeros, it couldn't escape from zero.

The moon's zeros went around and around and around but didn't point anywhere or go anywhere. They chased themselves in circles, snakes swallowing their own tails. The zeros closed themselves off and closed the entire universe out.

The moon's zeros stared at Earth. Cosmic zeros often infiltrated human hearts and made them ache with loneliness, made them long for their lives and the universe to add up to something. Humans looked at the moon and fell in love with goddesses but they were really seeing zeros. Humans sent their prayers and sacrifices to the moon but they were received by zeros. Human souls journeyed to the moon to find heaven but they found only zeros. Lovers enchanted by moonlight spoke of forever but it was the forever of zeros. Nations longed for the moon to make them feel superior but their rocket bells and landing pads only added more zeros to the cosmic scoreboard.

Yet it didn't take much to transform all those zeros into something, something enormous. It required only a "one." Not a seven, not twenty-seven, not a million, only a one. A one added to even three zeros turns them into 1,000. A one added to six zeros creates one million, which has a lot of value. A one has a lot of power. It is pointing somewhere, pointing upward, connecting Earth and cosmos. A one added to all the zeros on the moon transforms them into a massive number.

And now there was a One standing on the moon. Standing thin and straight, more vertical than any moon rock, denying the shapes of zeros, denying the chaos of rubble. An exclamation mark was standing on the moon. As it had come in for a landing, the One had maneuvered to avoid all those zeros, for it was incompatible with zeros, which would not have allowed its lander or itself to stand upright and point somewhere and not exclude the universe. Yet the One had risked becoming a zero to visit all these zeros, for humans had always been haunted by zeros.

The One stood there absorbing the zeros, absorbing them through the two zeros atop his face. His eyes were more geometrically perfect than any of the zeros on the moon, and more skilled. Through his eyes the moon zeros revealed that something lay beneath them, not nothing, something real and solid. The zeros existed only because they were revelations of the substantiality of the moon.

The debris of zeros upheld the astronaut. On Earth, such zeros had turned into eyes. Earth loomed in the darkness above him like a giant blue eye. To get a closer look, this eye had sent detachments of itself to the moon. It had counted down to zero, the big bang, and lifted off. The astronaut stood probing the zeros. He stood on the shoulders of giants, of both mystics and scientists who had forever contemplated how the universe could have hatched from a zero.

Earth loomed in the darkness above him, a giant blue zero.

35
GENESIS

The moon was oblivious. It didn't know what was happening, that life was walking upon it. The moon registered life's touch only as a compression of dust and as a slight vibration. Life's touch was so gentle compared with the touch of asteroids. The moon had long felt Earth's touch from afar, felt Earth's gravity controlling its rotation and occasionally luring a speck of dust to tumble half an inch. Now Earth's gravity in a more concentrated form was rearranging the dust in a far more sophisticated way, grasping it into Earth shapes, footprints that served a deeper effort to grasp the moon.

The moon had always been oblivious about itself. It had never wondered about its own existence, its origin, nature, relationship with the cosmos, or fate. It was a heavy mass of indifference. The sunsets and the stars and the strange Earth never stirred any sense of mystery, longing, or poetry. The worst asteroid impacts left the moon untouched. The greatest volcanic eruptions could not bring forth any deeper vitality.

But on Earth, matter had undertaken a greater adventure, one of making connections. It began with molecules making connections with one another, gradually better and larger and more enduring connections, eventually DNA and cells and larger creatures. Life made its first connection with the cosmos by binding sunlight with soil and water and air. Life sought out connections with the world outside it by growing filaments and sensors for feeling objects and energy, food and hazards and safety. Gradually life perceived the world as cycles of sunlight and darkness, cycles of weather, cycles of seasons, cycles of nutrients, to which life tried to match its own cycles of birth and growth, stasis and decay. Life watched the stars swarming overhead, saw patterns in their arrangements and motions, and tried to figure out its connection with them. Most of all, life saw the moon, strange and beautiful, changing yet reliable, epitomizing all the mysteries of the sky. Now life's connection-making molecules swarmed to find connections with the moon, to create stories about it—brain knots that tied together the loose and disobedient threads of the world. Life saw that the cosmos was full of forces that supported life or opposed life or were indifferent to life, and saw that actions resulted from intentions, so life tried to read the intentions and personalities behind those forces, including the moon. Life spun thousands of stories about the moon, thousands of gods and goddesses with their own lives and ways, and created thousands of ways of interacting with them, rituals for honoring or evading them. With its hands and songs and dances and churches, life reached toward the sky.

Life's long wondering at the moon eventually led it to make connections between metals and carbons and fluids and gasses, between miles of pipes and wires, giving them highly intricate and correct shapes that could rise from the ground and fly across space to the moon, allowing life to finally satisfy its curiosity, to connect itself physically with the moon.

On the way, for the first time, Earth saw itself, not just a very local patch of itself, not just the horizon seen from a mountaintop, but the whole of itself. All the mountains and other obstacles of living on Earth vanished into a smooth and tiny sphere. Earth finally saw itself in the context of a vast and dark cosmos, and compared itself with the chaotic moon.

Floating there between Earth and moon, humans were inspired to summon some poetry out of their store of creation stories. "In the beginning," the air brought from Earth rippled with structure. In many creation stories there was no real beginning to Earth or the universe, which seem to have been there forever in some form of chaos, only a beginning to order and life. "God created the heaven and the earth." Only one god was getting the credit for a moon inscribed with the names of numerous gods. "And the earth was without form, and void; and darkness was upon the face of the deep." And the moon still was without form, and void, a time-warp remnant of the chaos from which Earth had escaped and arisen. And God declared the creation to be good. This was once a radical idea, not a mixture of good and bad and indifference and mischievousness, but wholly good, born from omnipotent benevolence. The gods had been debating this idea for thousands of years, and now this apparent verdict was pronounced to a moon abandoned to chaos. Earth had journeyed far out of chaos, journeyed from a confusion of matter to a mere confusion of mind, and as a further extension of its journey, as a symbolic proof of it, in a metal sculpture of a cell, Earth had sent itself to the moon, back to the chaos from which it had arisen, to a mountaintop from which it could clearly measure the distance it had come and proclaim its journey to be good.

36
GODDESSES

The moon flowed onward, flowed through space, flowed through time, flowed through forms: into magma and solid rock, into mountains and rocks rolling down mountains, into volcanoes and lava flowing into lakes of lava, into craters and boulders, into faults and rilles, into many forms of dust.

Nearby, on Earth, other natural forces flowed onward: oceans and rivers, wind and clouds and rain. Rocks flowed into forms they never could become on the moon. Life arose and flowed from one form into another, into ocean forms and land forms, plant forms and animal forms, finally forms that wondered about flows and forms.

After four and a half billion years, the moon flowed into a new shape. It grew faces, female faces, tender and wise and strong. It grew a womb and breasts, rich with fertility and nurturance. It grew legs for walking upon the earth and arms for embracing it or pushing it away. It grew magical powers, the ability to bridge the gap between Earth and cosmos, between humans and natural forces. The ancient lunar magma flowed into eyes and glowed an eerie red. The craters blinked to the sunrise and widened to the night and opened to the sounds of wind and birds and crying. The mountains flowed with breath and the volcanoes with milk. The rilles became hair abraided with a crescent crown. The boulders became magnets for questions and templates for answers. The dust rose up and sang chants and performed rituals that summoned the powers of the moon and Earth.

The moon flowed into human eyes and into brains and was curved and convoluted and filtered into new shapes and glows and powers. The moon learned all the secrets of human life, including their gnawing questions about why they lived and had to try, why they suffered and died, questions about the cosmos and how they fit into it. The moon glowed with strange thoughts and feelings. The moon gained access to the human imagination, to this new incarnation of the universe's long search for order. The moon looked up and saw itself floating through the sky, saw a mystery and a beauty that symbolized the mystery and beauty of creation. The moon imagined many faces for itself, no longer a rock face but human-like faces, sometimes male faces but usually female faces, for the human mind could not resist imposing onto nature the gender identities that were so defining of its own realities. The moon imagined all sorts of personalities and motivations for itself, some good and some bad, some human-serving and some self-serving, some simply mysterious. The moon became a goddess, often more powerful and important than the sun.

The moon was only one of many natural forces that flowed into the human mind and turned into gods and goddesses. The ocean flowed in and solidified into Poseidon, whose bad temper became bad tempests. The clouds flowed in and became generous, granting the gift of rain. The sunlight flowed in and became Ra and Helios and Apollo and hundreds more, who often had to battle to keep light from being devoured by darkness. The earth flowed in and became Mother Earth, fertile and reliable even when cold and barren in winter. Animals flowed in and acquired spirits and voices and thus could participate in human lives.

The moon acquired numerous names: Diana, Selene, Ishtar, Luna, Hera, Levanah, Persephone, Anumati, Anu, Coyolxauhqui, Seda, Morgana, and thousands more. The moon grew thousands of faces, contradictory faces that embodied a contradictory universe and contradictory humans. The moon grew thousands of stories. The moon's face turned into masks, murals, pendants, and statues. The moon turned into round, stone observatories for tracking the moon. The moon grew massive temples, some with towers or rooms or steps for every day of the lunar cycle. The moon became crystal balls, candle flames, torches, jewelry, and white dresses. The moon spoke as poems, chants, songs, and prayers. The moon spoke through priests in crowded temple ceremonies and through a moon-circle of women in the forest. The moon inspired all kinds of ceremonies that brought blessings, ceremonies for fertility, planting, harvests, marriage, pregnancy, birth, healing, longevity, death, the new year, happiness. The moon became round dances and round ceremonial cakes. The moon brought menstrual cycles in women and brought out the werewolf in men. The moon became the abode of the dead, or the eye of god. The moon swelled steadily, supernaturally pregnant, and gave birth to the entire cosmos.

In one African story, when the Great Spirit created Moon, Moon wanted to live on Earth. The Great Spirit warned Moon that Earth was barren and that he could not survive there for long, but Moon was determined, so the Great Spirit placed him upon Earth. Moon was dismayed to find no life there, not even grasses or bugs, only rocks and dust. The Great Spirit felt sorry for Moon and sent him a wife, Morningstar, and together they filled Earth with life.

Step by step, as Apollo flowed upon the moon, the moon welcomed itself home.

Above all, the moon goddesses embodied mystery. The goddesses were born from the womb of mystery, from a senseless universe determined to make sense of itself. The mystery of the cosmos became incarnate, turning immaculate space into a body bright and moving and bleeding, able to sympathize with and offer salvation to pure space exiled into human corporality. The faceless mystery gained faces through which humans could recognize it. The nameless mystery gained names through which humans could address it. The untouchable mystery gained hands through which humans could feel it. The mindless mystery gained minds with which humans could communicate and negotiate. The negative mystery might still be malicious but at least it now offered explanations. The glorious mystery might now be thanked in person.

From many sources, many kinds of mystery flowed into the moon goddesses.

From the Big Bang came the mystery of why there were any moons at all. From supernovae came the Shiva mystery of intermingled creation and destruction. From red umbilical nebulae came the mystery of gestation and birth. From stars came the mystery of time and endurance. From myriad solar systems came the mystery of order and chaos, of gravitational necessity and formational chance. From ice moons sparkling with sunlight came the mystery of crystalline structure and irrepressible energy. From tides pushing grains of sand up the beach and from gravity pulling them back down came the mystery of moons rolling like grains of sand. From deserts came the mystery of water. From dead planets and moons came the mystery of cells. From the flight of a luna moth came the mystery of a world transformed. From the brown earth and warm sensuous flesh came the mystery of fertility. From a crying baby came the mystery of pain and disease and injustice. From the final indignant breath of animals that had done their best to obey their cosmic orders to live came the mystery of death. From the fall of an apple came the mystery of human wonder.

All of this mystery spun itself into the faces and

forms of moon goddesses. With eyes intense with mystery she looked upon Earth and saw humans intense with mystery, awed by mystery, tortured by mystery, fighting one another over mystery, dying over the mystery of living. Humans looked at the moon and saw the imperfect fertility of females and fields, saw the secrets of death and resurrection, and they made altars and dances to the moon to urge her on. The moon goddess was, in her own way, a space probe for exploring and defining the cosmos. The same mystery that created her helped to draw humans to the moon, and though they were not seeing Diana they did have enough sense of mythos to call themselves Apollo, and they did sense the mystery of the moon. But they also saw that humans had imposed their identities, their hopes, their narcissism, upon the moon, demanding that it serve fertility and resurrection when in truth it was a grey wasteland. In truth, it was an even deeper mystery.

37
CIRCLES OF STONE

With millions of stone circles, the moon monitored the circlings of the solar system. Crater rims, with their peaks, boulders, notches, and slopes, made good horizons for showing the planets rising at shifting locations every night, every very long lunar night, shifting steadily onward, then steadily backward, over and over through millions of years. Craters could predict exactly where the sun would arise, and exactly when Earth would eclipse the sun. The sun drew long light daggers and long shadows across the ground, pointing at a boulder here or a distant crater there, pointing out that the moon belonged to a clockwork cosmos. Millions of moonclocks ticked onward. Obliviously.

Through a circle of stones, the full moon arose and climbed into the sky. The stones monitored the moon's motions through the days, the seasons, the years. The stones could foretell when the moon would eclipse the sun. The stones also marked the sun's seasonal cycles across the horizon and through the sky. The stones watched Venus and Jupiter and Mars wandering.

This stone circle wasn't nearly as old as moon craters. These stones were different, still rough but also smoothed. They sat on flat land, not perched atop craters. They had not been raised by cosmic violence. Like moon stones, they were oblivious of the time and the motions they counted. But through them, time was lived, the motions and phases of the moon were lived. Through this portal, the cosmos was translated into longing and gratitude, and stone became hoes to seed the fertile earth.

From afar, the circles of moon stones studied the Earth stones. Twins at birth, the moon was unable to grasp the distance that had opened up between it and Earth, or Earth's strange gravity.

38
NAMES

In the beginning, there were no names. The universe was content to be itself, pure and uninterpreted by symbols. Matter itself was all that mattered. God, too, if indeed God created the universe, was content to be pure being for billions of years; perhaps God waited so long to create life because God was enjoying namelessness, enjoying what really mattered, and did not want to be distracted by the reputation-muddling demands of needy little creatures. It was humans who started the rumor that God had named the light "day" and the darkness "night" and the lesser, night-ruling light the "moon."

Humans looked upon the moon and projected their own lives onto it. They saw the moon as thousands of goddesses and gods, with thousands of faces, personalities, and names. They performed rituals and built temples to coax the moon into speaking their names and answering their needs. Even when they saw the moon as another physical world, they saw it through water-planet eyes, saw the Sea of Fertility, the Ocean of Storms, and the Sea of Tranquility. In novels and movie theaters they projected lunar creatures and dramas. When humans named lunar mountain ranges they named them for Earth mountain ranges. When humans named lunar craters they didn't name them for Earth craters or volcanoes or canyons, but for humans, for hundreds of scientists, explorers, mathematicians, philosophers, inventors, and a handful of visionary writers like Jules Verne. Among the most visible craters were Copernicus, Kepler, Tyco, Archimedes, and Plato. Astronomer Percival Lowell was awarded a crater even though his most famous accomplishment was projecting onto Mars imaginary canals and Martians. Biologists like Carl Linnaeus and George Washington Carver got craters on which nothing could grow. Craters were named for two German physicists for whom national narcissism defined morality: Fritz Haber, the father of chemical warfare in World War One and the Nazi gas chambers; and Johannes Stark, who led the Nazi campaign against "Jewish physics" and Jewish physicists like Einstein. But at least Einstein got a lunar crater.

Now astronauts who would soon have craters named for themselves walked upon water, upon the Sea of Tranquility, but they found it was only dry dust. If it held tranquility it was not the peace of a city park but the tranquility of a cemetery. They were the only faces there, besides faces of rock. They were the only gods, and cautious and fleeting gods. The craters did not look anything like people. The real craters had no names and no fame written on them. The craters did not welcome the walking craters. The only human shapes the astronauts saw upon this ground were their own shadows. Their shadows were projected in an empty theater that did not acknowledge humans, a theater to which God retreated to enjoy primordial namelessness.

If the astronauts were walking in crater Plato, their shadows would have been the shadows in Plato's cave, misinterpretations of reality. If the astronauts were walking in crater Xenophanes, who lived a century before Plato and satirized Greek anthropomorphic religion as a mere human projection, saying that if horses and cows had religions they would imagine gods that were horses and cows, the astronauts would have seen an astronaut god and themselves as his finger puppets. If the astronauts were walking in crater Pavlov, they would have heard the incessant bells of human society and human churches training humans to expect rewards. If the astronauts were walking in crater Giordano Bruno, their shadows would have lurched from the bonfires of human narcissism. If the astronauts were walking in craters David Hume or Niels Bohr, they might have seen no shadows at all.

The astronauts noticed their shadows imitating their every move. Yet when the astronauts spoke the names of things or their own names, their moon shadows were entirely silent.

39
THE CLIMB

The astronauts were running out of the order they had brought from Earth; it was draining into the chaos of the moon. They were running low on oxygen, water, food, battery power, alertness, and especially time. To the timeless moon they had brought far too little time.

One astronaut returned to the lander and began climbing its ladder. The moon resisted. Even in low gravity, climbing was hard work. The lander's compressible legs had not compressed as much as expected, so the ladder's lowest rung was too high off the ground, requiring a good, accurate jump to reach it. With bulky spacesuit and gloves and boots, grasping the rungs was awkward. The astronaut reached the door and unrolled a tether for hauling up the boxes of moon rocks, but this too proved an awkward struggle. The moon resisted, as chaos had always resisted the order trying to arise from it. With the boxes secured, the second astronaut jumped and climbed into the lander. Chaos resisted, as it had resisted molecules climbing into DNA, cells climbing into multicellular life, fish climbing onto shore, and primates climbing into trees.

When the rocket ignited and the lander lifted off, the moon did its best to overrule it, to draw it back to chaos and make it crash.

40
NUNKI

It was an ordinary star, one of more than 200 billion stars in the galaxy. Born from a nebula with many other stars, it traveled with them for awhile but gradually drifted apart, forgetting its siblings. It traveled in the great circular flow of the galaxy but also in local directions, propelled by its own birth momentum and by the pull of neighbor stars or wandering stars. It wandered onward for eons, indistinguishable from millions of other wandering stars.

This star never knew that one day it was singled out by creatures who were lost in space, singled out to help them find their location and their way. From 228 light-years away astronauts pointed a sextant-telescope at this star and translated it into a precise number. They had planned to use two other stars but were surprised these stars were below the horizon. They tried other stars, but these stars too were out of view. So they focused on a star that out of 200 billion nameless stars had been given a name: Nunki. Oblivious of its own location or of the existence of Earth, Nunki told the astronauts on the moon exactly where they were (for they had landed a few miles off target) and told their machine the precise trajectory it would need to follow to climb into the correct orbit. Burning and wandering and oblivious, Nunki too was lost in space, far more lost than the astronauts would ever be. Nunki would never know it had entered into an ancient yet always new drama of lostness and discovery.

In looking at Nunki, the astronauts were looking very close toward the center of the galaxy, at a black hole that held, in a space about the size of Earth's solar system, the mass of some four million suns, a black hole that poured out massive amounts of energy (invisible to human eyes) as it consumed stars and planets and moons. The astronauts were also looking straight at several nebulae—including the Lagoon Nebula and the Swan Nebula—where stars and planets and moons were bring born. Standing between creation and destruction, between unimaginable power and emptiness and mystery, the astronauts imagined they had located themselves with one simple number. Nunki never acknowledged them. Even Nunki, even the moon, even Earth, even the astronauts' own faces were invisible to human eyes.

This was not the first time Nunki was cast in a role in a human drama. For thousands of years humans had seen Nunki and imagined shapes and stories for it. Nunki is the second brightest star in Sagittarius, a constellation recorded in the stones of Babylonian tablets and monuments and in the early zodiacs of Egypt and India. Sagittarius was passed from one society to another, remaining consistent while societies crumbled. The Greeks envisioned Sagittarius as a centaur holding a bow and aiming an arrow; Nunki was part of the centaur's outstretched arm or sometimes part of the shaft of the arrow. Sagittarius was a hunter and sometimes a warrior and took part in the lives and conflicts of the gods, sometimes killing or being killed. Nunki never imagined the flesh, the lives, the passions, or the godly powers ascribed to it. Nor did humans imagine that Nunki was another sun, and not an ordinary sun but a sun 3300 times brighter than our sun, rotating 100 times faster and living a far shorter life. Nunki and humans lived in nearly separate universes. But now they had reached out across space and made contact, hand to lightbeam, as humans sought to comprehend their place in a strange universe of giant stars and black holes and dusty moons.

41
SLEEP

The moon rocks rested, as they had rested for eons, not moving or wanting, barely decaying. But now they rested inside two metal boxes that gave them shape as nothing before had given them shape, gave them boundaries and depth and stacking and rounded corners. The boxes were meant to keep two worlds separate, to keep the rocks guarded by lunar vacuum and to keep all Earthness outside.

Removed from their longtime natural habitat, placed inside the lander, the moon rocks were surrounded by strange complexities, by geometries, materials, arrangements, open spaces, and purposes; by machinery, tubes, wires, switches, gauges, buttons, and symbols; by sounds, lights, and motions. But of course the moon rocks never noticed; they continued their sleep of eons.

After awhile the motions and sounds and lights settled down. The moonlight coming through the window was eclipsed by curtain. The two humans settled onto the floor and curled up and became motionless, mostly, and closed their eyes to disconnect the orbits within their brains from the orbits and chaos outside. Normally humans relied on the turning Earth to turn off the lights, to turn them into the realm of darkness and moonlight where light would not stimulate a light-hungry, light-survival animal too much. Thus humans associated moonlight with sleep, dreams, and a gentler and more whimsical consciousness. But these two humans had left the protection of Earth and turned the sprite moonlight into a glow far too bright, even through their window shades.

The humans lay motionless, almost like the moon

rocks, yet inside them motions continued, some slowing yet others picking up. How could you explain sleep to moon rocks? It was not the sleep of rocks. Brains and muscles were abuzz with molecules performing rest, removing tiredness and restoring energy. The motions of the day, of landing on and walking on the moon, were being erased. Moon scenes were being cataloged into memories. The joys and worries and details were fading into a matte painting. The brain was erasing the intensity of moonlight and moon touch, just as humans always insisted on blotting away the intensity of existence itself. The brain flickered with images and tried to make sense and stories of them, just as it always tried to make sense of the flickering stars and the flickering moonlight and the flickering mysteries of life and death. The moon rocks waited patiently, waited with memories merely geological, waited to help a dreaming planet become more awake.

Then, oddly, upon the face of one of the humans, a light appeared and grew wider and brighter and bluer. This spotlight was aimed straight at his eyes. He tried to turn away but there was no space. He opened his eyes and squinted and tried to identify the source: it was the eyepiece of the telescope used for navigation and docking. Through it, lined up perfectly, focused just right, Earth was peering straight at him. A high-tech Stonehenge.

As if through a microscope, Earth was examining him; the blue ocean was probing the blue eyes it had given birth to. What was it seeing? What was it asking the detachment of itself that had traveled so far and seen so much? If Earth was looking through his eyes and brain, it was seeing the primordial desolation it too had once been. It was seeing the footprints in the dust. And it saw a man unable to withstand an angel's gaze incandescent with reality, a man trying to turn away, to close his eyes, to retreat into dreams, unable to sleep.

42
MOONDUST

The lander held not just moon rocks in their boxes but also lots of moondust, roaming free.

With every step the astronauts had taken on the lunar surface, dust had sprayed outward and landed on the opposite boot and leg and stayed there, slowly turning the white spacesuits grey. As the astronauts handled moon rocks, their gloves and arms too turned grey. When the astronauts climbed back into the lander they carried lots of dust with them and smudged it onto everything they touched. As air refilled the cabin, it lifted dust off the floor and walls and spacesuits, billowing it throughout the cabin, painting everything, tucking it into every switch. The astronauts inhaled moondust. The astronauts might try to clean off the dust but plenty would remain to billow into the command module when they docked, so the astronauts would be breathing moondust all the way back to Earth.

From the nose, the dust flowed down the throat and into the lungs and stomach. From drinking tubes and from the air, the dust touched lips and was swallowed. The moondust roamed through the bizarre, moist, sculpted, pulsing, crowded corridors of life. It flowed into a maze for testing molecules and triggering the shapes of memories: the memory it lit up was the scent of gunpowder. Through membranes that inspected it and accepted some of it, it was loaded into blood cells and sped through the heart and around the body, energizing cells and muscles, building new DNA and new cells, helping the astronauts move fingers and breathe in more moondust. The moondust was taught in seconds what Earth had taken billions of years to learn. The moondust curled into fingerprints, into the myriad and terribly complicated identities of Earth. The moondust became a nose and smelled gunpowder and learned that the life of Earth did not appreciate itself and enjoyed destroying itself.

The moondust was guided into eyes and sewn into molecules and given shapes curving and clear, and taught how to capture and focus light and translate it into another kind of light. Moondust that had sparkled only by discarding sunlight back into space now sparkled with insight and joy. But now the moondust was separated from its fellow moondust by windows and visors that filtered down the sunlight, for eyes could not withstand naked sunlight; this was the moondust's first lesson in the delicacy of life, in the concept of pain and dysfunction. This entire human body was thoroughly defended against alien forces.

The moondust traveled through the maze of the brain, searching for forms and forces that needed reinforcements. It traveled through forests of memories, through the bases of many abilities, through clocks and messages keeping the heart beating and sleep recurring, through monitors of the inside and outside worlds, through the fires of consciousness. It was initiated into the deepest secrets of the brain, secrets the brain itself did not know consciously. It began interviewing the light streaming in from the eyes and being enlightened by it. It learned the secrets of itself.

The moondust looked out the window at the moondust, seeing its color, texture, shapes, and shapelessness. Hours later, as the moondust shot into the moon sky, it saw the moon diminish and yet expand into larger patterns, into roundness, into a grey ball floating in space. The moondust saw what it had been and what it had been doing all these eons, doing very little except crumbling from rocks and sitting and waiting.

The moondust saw itself through the filter of human

experience and thus did not really recognize itself, saw itself as strange and alien. The moondust looked through biological eyes that loved order and regarded dust as disorder, a contamination of life's abode, a threat to health. Humans devoted a great deal of time and effort and money, a great deal of water and machinery and substances, to cleaning dust off their bodies and hair and clothes, off their furniture and appliances and floors and cars. The lunar lander had been constructed in a super-clean facility with all sorts of devices, clothing, and procedures for keeping out dust and dirt, which might interfere with electronic systems. Could humans look at moondust without feeling that it was chaos, worthless, unfriendly? Human minds were programmed to scan the environment for shapes useful to life, and there were no obvious such shapes on the moon. Human minds liked colors and often used the word "grey" for drab or depressed.

The moondust had to look at itself through centuries of metaphors about dust. Humans liked to frighten themselves with dust. Ashes to ashes, dust to dust, everything ended up as dust: human bodies, even the bodies of kings; human cities, even the palaces of kings; human hopes, even empires with all their pride and pomp.

Yet perhaps the moondust could help teach humans to see the cosmos not with merely biological eyes but with astronomical eyes, to see dust, both moondust and interstellar dust, as rich with order and pregnant with possibilities. Moondust was a very sophisticated substance, holding the whole of cosmic evolution, the energies and creativity and architectures of the Big Bang and stars and supernovas, and capable of rising into further order. "Dust to dust" does mean that dust can rise into the forms of life. Yet humans spoiled the cosmos for themselves by perceiving it through the needs of life, worrying about eating it or being eaten by it, defining the moon by the tolerance of human skin for heat and cold. If humans could see the universe through moondust eyes, they might find it—and themselves—more beautiful and meaningful.

The moondust was initiated into the ways of living bodies, their solidity yet fluidity, their pulsing with air and water and solids, their limbs and motions, their hunger for food and motion and reproduction, their kaleidoscope of feelings, their lonely boundaries between individuals, their weird rules of social behavior, their divergent and obsessive identities, their condemnation to pain and death. The moondust was initiated into human history and culture, learning about Egyptian pyramids and Roman armies and Greek temples to the moon. It learned of despairing and yearning prophets retreating to the deserts and mountains to seek a glimpse of god. It learned that the moon had become gods and goddesses with numerous names and shapes to regulate the cosmos and give sense to human life. It learned how the moon's laws had arrested Isaac Newton and curved Albert Einstein's brain. It saw painters and poets turning the moon into symbols of loveliness and longing. The moondust was god become flesh, come down to Earth, sacrificing itself so that humans might live.

43
THE CHOSEN

Surrounding the lander were hundreds of rocks, not boulders, for the biologicals had landed here to avoid too much geological chaos. The rocks might seem to be randomly scattered but there were subtle patterns to them, lines and clusters and sizes and angles, patterns that mapped out the shapes and energies and trajectories of ancient asteroids and the strengths of the lunar crust. The rocks looked essentially alike, and alike rocks all over the moon, but now a few of them were chosen for a strange fate. They were chosen partly at random, the randomness of whether the lander had landed right here or two hundred feet away, the randomness of where an astronaut walked and turned and looked, the randomness of how primate evolution had sized and shaped the human hand, the randomness of whether an astronaut was right-handed or left-handed, the randomness of what looked good to different astronauts. But the rocks were also chosen because their color or sparkle hinted at geological secrets within, or fit an aesthetic impulse humans had picked up in evolution or in art museums. The chosen few were packed into bags and boxes and propelled by a strange force, muscle power, into the lander. They were forever separated from their buddies. They had been given a new identity: after being the only rocks in their known universe, they had become moon rocks, distinct from Earth rocks or Mars rocks. After eons of being anonymous, they would now become famous, great anomalies. They had been appointed ambassadors from the moon to Earth, carriers of secret stories, silent oracles. They held a cryptic language spelled in atoms and molecules, written by time, written by lava and asteroids, written by a nebula and gravity and the entire solar system, written by Earth itself. The rocks would tell Earth its own secrets. The rocks had sat there remembering, while Earth was erasing its memories.

The rocks sat inside the lander for many hours while the astronauts slept, and then the rocks leaped off the moon, vibrating intensely, riding an anti-asteroid, riding an anti-gravity force, riding the force of consciousness. The rocks flew away from the moon.

44
RENDEZVOUS

The astronauts were filled with rendezvous. Throughout their bodies millions of atoms and molecules were rendezvousing every minute, guided by intricate and precise systems for bringing them together and snapping them into new molecules. Ignorant, serf carbon atoms swarmed onto long templates and emerged as genius molecules that quickly filled the gaps in DNA and ruled the whole process. Clusters of ability flowed into cells and organs and were assigned larger purposes. Everything rendezvoused into a symphony of shapes and energies and timings, into life.

Many of the raw materials life was using had been fitted together outside of life or before there was life. Earth's geological and hydrological forces had gathered, purified, bonded, and distributed many elements. The oceans were a banquet of water and salt and organic molecules for life to thrive on. Even the moon, before its inner fires died out, had been processing elements into larger patterns. Other planets and moons continued doing so, in their varied ways. The planets and moons were born from greater, astronomical rendezvous, spinoffs of the sun's birth from its nebula and from the galactic merry-go-rendezvous. Inside the sun, atoms rendezvoused violently and snapped together, a continuation of the meeting with which the universe began.

And now as a continuation of all that rendezvousing, the lander rose from the moon, no longer a lander because it had left its legs on the moon as an abstract sculpture of the Earth legs that had touched the moon. The ascent stage had fired its rocket at a precise moment, with a precise force, in a precise direction, for a precise duration, and it slid into the same orbit as the waiting command module and approached it. The command module fired its thrusters and aimed for the black circle atop the lander, the perfectly smooth, unlunar crater, and slid its cone into it until they were snug and a circle of bolts fired and bonded the two spacecraft back into one.

This rendezvous was a continuation of the precise trajectories and meetings that occurred throughout the mission. When the spacecraft had left Earth it wasn't aimed at the moon but at a spot in empty space nearly 200,000 miles away from the moon, a spot towards which the moon was racing, and they arrived not at exactly the same moment but with the spacecraft firing its rocket to ease itself into a cozy orbit. When the lander headed for the lunar surface it followed another precise progression. To get back to Earth the command module had to aim itself precisely and burn its rocket with second-accuracy, and to reenter Earth's atmosphere (which it would at an unavoidable seven miles per second after falling from the moon), to avoid either burning up or skipping back into space forever, it had an extremely small margin for error.

All the Apollo rendezvous were a magnification of the merging of carbon atoms into DNA and cells. They were the continuation of fourteen billion years of matter meeting itself. They were a new kind of meeting, the meeting of Earth and moon, of order and chaos, of present life and the ghosts of universes past. They were the universe meeting itself in a large, round, white mirror.

45
HEXAGONS

Sunlight incarnate, the bees swarmed over the meadow of yellow flowers, loading themselves with solidified sunheat. Flying toward home, they were buffeted by the wind yet they constantly adjusted back to a bullet trajectory mapped by the sun. They flew over water on which floated small rafts of bubbles, bubbles that the laws of physics, the forces of surface tension, had arranged into hexagons. The hexagons sparkled with sunlight. The bees passed flies and they saw one another with eyes made of thousands of hexagonal facets, arranged with more geometrical precision than the bubbles. The bees flew over turtles with hexagonal segments in their shells. The bees flew over water that in winter had fallen from the sky and crystalized into extremely elaborate and unique hexagons, sparkling as they fell, sparkling with the warmth that would soon dissolve them. The bees flew under the long overhang of a volcanic cliff, with thousands of black hexagonal columns, the shape into which basalt had cracked as it was cooling and hardening, the shape that most effectively released its heat and tension.

The bees flew to their nest under the cliff, their honeycomb of thousands of hexagons, hexagons far more symmetrical than those in water bubbles if not as perfect as those in insect eyes. The bees unloaded their honey. On the edge of the honeycomb some bees worked at repairing and expanding it, secreting soft wax whose surface tension helped bend it into hexagons. But surface tension wasn't used by paper wasps, who digest plant fibers and secrete and weave it into a paper-like fabric, into nests as hexagonal as bee honeycombs. Bees and wasps find hexagons logical because hexagons take the least work and the least material to form the most chambers for storing food and eggs and for raising young.

As the rising sun reached into the cliff, the honeycomb gleamed, as if with recognition of the source of its stored energy, as if it were the sun's own eye.

One morning the bees saw a second sun appear, moving steadily and growing brighter, a flame streaking behind it. A honeycomb was falling from the sky, shedding its honey. A turtle was falling from the sky, holding out its shell to defend itself. Bees and turtles and flowers and humans could live off the sun's heat only because space and a thick atmosphere filtered the heat into a softness their delicate bodies could withstand and use. Yet the atmosphere would also filter out human bodies trying to pass through it, turning their speed into heat and disintegration. Out of eons of generating hexagons, out of lava using hexagons to dissipate heat, out of turtles using hexagons to block heat and teeth, out of bees using hexagons to store heat, Earth had now generated a shell with a third of a million hexagons packed with "honey," with resins that would heat up and ablate away and carry away thousands of degrees of heat.

The bees and turtles watched their own hexagons coming home. When the spacecraft hit the sea it sent out waves of bubbles that had known the secrets of the cosmos all along.

46
REENTRY

The moon rocks sped toward Earth, sped at 25,000 miles per hour, sped like the billions of asteroids and comets that had sped toward Earth, like the asteroids that had sped toward the moon and smashed its crust into these rocks. The moon rocks sped toward Earth but began encountering a resistance that asteroids had never felt at the moon, the immune system with which Earth defends itself against foreign bodies. The moon rocks began to generate heat, or at least they contributed a bit to the weight that generated heat around the metal capsule in which they rode. The heat built up, up to 5,000 degrees Fahrenheit, creating a fireball with a long, pulsing tail. Such heat had broken up most of the asteroids and comets that approached Earth, burned them into cinders that had fertilized the land and sea, yet these moon rocks had learned the secret combination to Earth's lock. They were protected in just the right shape and substance, pointed at just the right angle, and followed a narrow tunnel through destruction. The heatshield beneath them melted and streamed away, carrying heat away with it, exposing new layers to take the heat. The fireball glowed green and yellow and purple, changing colors as the capsule sped up and burned hotter. The fireball streaked through a thousand miles of air, streaked through the night sky, heading for the boundary between night and day. The fireball generated a tsunami of air, a sonic boom. The moon rocks and astronauts felt six times the force of normal gravity, as if paying for the one-sixth gravity of the moon. Miles above the ocean, parachutes flowered out and slowed the capsule. The moon rocks swayed back and forth in the wind. They felt a sudden stop and rebound, and then they swayed back and forth in the waves, in the motherly tidal summons of the moon, in the welcome of a nurturance it had never imagined. The charred surface of the heat-shield began dissolving into the waters and drifting away to be merged into turtles who would wonder at the moon they had once orbited.

These moon rocks had once ruled the tides that now tossed them. In all, 842 pounds of moon rocks would come to Earth, subtracting from the lunar throne yet being traded for far more Earth matter left on the moon, which would make the tides slightly more Earthly, more human, more self-inspired. It was a reenactment of the trading of matter that had formed the moon and left large amounts of Earth within it, a still unrecognized trading that these moon rocks would reveal. In their molecules the moon rocks said: *we were never really moon rocks; we were Earth rocks all along, exiled and forgotten, left out as Earth journeyed onward, but now we have come home.*

If they were free to do so, these moon rocks would join the charred heatshield in dissolving into the ocean and flowing away to join the tides and rains and turtles, their moon identities erased, but humans would enforce their identities by keeping them sealed in vacuums in transport boxes, laboratories, and museum displays.

Yet a bit of the irrepressible moon did escape. When the hatch was opened, the capsule's air rushed out and the outside air rushed in, scouring every nook hiding moon dust, blowing it out the hatch and onto the waters, on which it swirled away to join the swirls of turtle motions and turtle brains. A few weeks from now children swimming along beaches would be floating on moondust.

For eons these moon rocks had sparkled with Earth-light as Earth glowed in their sky, growing and shrinking, Cheshire smiling at them, as over eons Earth's face had changed from red to blue and white and green.

Now the moon rocks discovered that Earth had pix-ilated into many smaller faces. In museums, the faces crowded around the moon rocks, their ocean-filled eyes sparkling with moonlight, their inner tides responding. The faces came and went, faces of many sizes, shapes, colors, furs, and decorations, all of them cells of Earth's cosmic face. The moon rocks would never see them, for that was the identity of moon rocks. But Earth saw the moon rocks, sought something in them, something that had urged Earth to reach so far. Into the moon rocks, opaque and yet lucid, Earth gazed to try to glimpse it-self, to see its own true identity at last. Earth stared, and the moon rocks wavered, and images congealed. Only in formlessness could Earth see its own forms; only in still-ness could it see its own journey; only in paralysis could it see its own abilities; only in greyness could it see its own colors; only in blindness could it see its own vision; only in silence could it hear its own music; only in obliviousness could it see its own mind; only in death could it see its own life.

And then one day came a face that might have seemed familiar if the moon rock could remember, a face that had appeared upon the moon, an apparition, and looked at this rock and liked it and picked it up and carried it to Earth. The astronaut stared at the moon rock and re-membered the moon, the feel of it beneath his boots, the silence that magnified his own breathing, the shapes and textures, the strangeness of it, the deep reality of it.

That night, when the museum lights were turned off, the moon rock was engulfed in darkness, until a vague glow appeared and grew, vaguely lighting the moon rock, casting a vague moon shadow. The light grew. Through a distant window appeared the moon. It reminded the moon rock of how it had glowed on the moon, how it had been the moon. Mother moon called out to her child, whispering deep secrets to tell to Earth.

EPILOGUE:
THE ECLIPSE

The day before the eclipse, I walked through a museum full of skulls. They looked at me with empty eye sockets and motionless mouths and tried to tell me about their lives.

A brontotheres, rhino-like, tromped through the mud of a dense, warm, humid, semitropical rainforest, tugging leaves and vines off trees, while crocodiles edged out of his way.

Tapirs wandered through the forest at night, recognizing leaves by moonlight, a moon that grew and shrank reliably for the next forty million years as continents and habitats shifted less predictably, forcing tapirs, who refused to change form, to migrate to new areas. This tapir skull could have belonged to today's tapirs.

A small herbivore looked at me from a cracked skull that had been stepped on and crushed by a larger mammal, perhaps a four-toed horse or an ancestral bear whose skulls had come through more intact, if missing some teeth.

The ancestors of today's dogs and cats told me about living in a hardwood forest, its precisely fossilized leaves and nuts displayed nearby. The dogs and cats had done just fine without humans.

A camel skull and rabbit skull agreed about how forests had gradually turned into grasslands.

Several skulls were tinted green, the same color that banded the landscape outside, telling me of the minerals that fell from volcanoes and of the infiltrations of time.

The eye sockets of a beaver had watched her newborn beavers playing on an endless spring afternoon.

A *bergagnathus aterosseus* mouse skull complained that it didn't recognize this name it had been given, that her true name had been a scent and a face and a peep.

Over some forty million years these skulls had watched the sun rise and set and the moon come and go and the seasons change, and they had recognized that the world held a reliable order. Some of these skulls had one day noticed the daytime sky darkening unnaturally and the sun going out, and they had watched in awe and perhaps run to hide from a shadow more ominous than the dinosaur shadows still imprinted in their ancient mammal memories.

I was going to watch this eclipse from the John Day fossil beds, one of the world's richest fossil deposits, revealing some forty million years of mammal forms and evolution, revealing mammals rising into shapes ever more diverse and agile and dexterous, more smart and sociable and nurturing. I was here because this section of central, drylands Oregon offered the best chance for clear skies along the eclipse's route across the entire North American continent, and because it was public land, a National Monument, offering better access. By coincidence, I was also being offered one of the world's clearest views into biological time and forces. The fossils were here because this was volcano country, where over millions of years volcanoes had pumped out a series of ash clouds and mud flows that had covered the land deeply and buried its life and preserved their forms, often with amazing details. The volcanoes had left this a badlands of odd colors and tilted stripes and erosional oddities, often with little vegetation to hide them.

In coming here, I was fitting into a cycle of the cosmos. This eclipse was essentially a continuation of the eclipse I had seen eighteen years, eleven days, and eight hours before at Verdun. I had not kept track of that time, but the solar system had. Eclipses are embedded in long, evolving cycles of eclipses called Saros cycles, shadow puppet shows cast by the intricate motions of the sun, Earth, and moon. The Saros cycle that generated these two eclipses would last 1,370.3 years and generate forty-one total eclipses and thirty-four partial eclipses. This Saros cycle began in 1639 near the North Pole, and each subsequent eclipse moved southward and westward, at first only partial eclipses, but in 1927 this cycle had generated its first total eclipse. As the cycle continued, its eclipses would migrate toward the equator and become longer, peaking with a seven minute, twelve-second eclipse in 2522. Then its eclipses would migrate towards the South Pole, again becoming only partial eclipses, and go extinct in 3009. Other Saros cycles started at the South Pole and migrated northward. About forty Saros cycles were running at the same time, crisscrossing each other's paths, generating several total eclipses somewhere on Earth every year, generating narrow, curving tracks sometimes thousands of miles long, writing, with shadow pens, precise mathematical equations onto Earth, the signature of cosmic order. Because each eclipse in a Saros cycle occurred a third of Earth farther westward, I had come a third of the way around the planet to meet this cycle again.

Eclipse cycles and sky cycles were only the most visible manifestations of cosmic order. They were fulfillments of the ordering impulses that sprang from the Big Bang. They were magnifications of the cyclings of electrons in atoms and the reliable rules and structures in the heart of atoms. They were elaborations of the gravity that had saved matter from dispersing forever into nothing but a vast haze. They were smaller gears of the turning galaxy. The sun with all its chaos still held great order, and the moon, hatched from suns, held, even in its craters, portraits of the solar system. On Earth, cosmic order had not been frozen as craters but had rolled onward into cells and footprints. Eclipse cycles and sky cycles had rolled into heartbeats, shaped themselves into all the fossil shapes around me now.

The Native Americans who had lived here must have recognized the shapes of bones and skulls and teeth in the ground and known that this earth was full of life and magic and connections with themselves. In 1865 Thomas Condon, a Congregational minister from a town a hundred miles away, came to these fossil beds, which he named for the nearby John Day River, and recognized their great value. As a youth, Condon had been fascinated by fossils, and lately he was enthralled by the rapidly expanding science of paleontology. He excavated fossils and sent them to some of the leading scientists of his time, and soon they were coming here to do their own excavations and to fill their museums, to illustrate the new idea that Earth was very old and full of changes. Condon arrived at the John Day fossil beds only six years after Charles Darwin had proposed that life too was very old and full of changes, and his fossils became chess pieces in the worldwide debate over evolution.

For many Christians, Darwin triggered a crisis of faith, for life and humans no longer required God to create them, and creation no longer fit religion's moral compass. Thomas Condon felt the tension between his two passions and insisted that science was simply a further revelation of God. He adopted William Paley's logic that a well-made watch proved the existence of a skilled watchmaker, and he went further and accepted geological and biological evolution as instruments of divine creativity. Condon tried to expand God to fit expanding time and Earth and cosmos. In one public lecture he held that an eclipse that had passed through Oregon a generation before was like the ticking of Paley's watch. At the entrance to the park museum, named for Thomas Condon, the National Park Service offers a Condon quote reassuring visitors that fossils are not atheistic.

It had always been reasonable that when humans saw the order of the cosmos they would reason that order-making agents were behind it all, and since by far the most adept order-making agent they knew was the one with two hands and a creative mind and bodily desires and social necessities, it was not surprising they filled the cosmos with gods somewhat like themselves, gods to whom the order or disorder of the cosmos could be appealed. It was not surprising that when an eclipse arrived it was a shocking violation of the natural and supernatural order, an outbreak of chaos, a failure of the gods, an attack upon the gods, or a dark message from the gods to humans, prompting humans to react with fear and pleading.

For me too an eclipse revealed the gods that ran the universe, but for me the eclipse represented order, wheel within wheel of order, trustworthy and creative and nurturing order, the order that had generated life. Humans had excavated an order that went far deeper than the events on the surface of animal perception, deeper than animal reasoning. The eclipsed sun was a portal, a black hole, into the universe's secret heart. Its seconds were ticked away by a clock 1,370.3 years old, which itself was but one tick of the universe's heartbeat. The moon in all its crudeness was still a masterpiece of roundness and motion. Sun and moon aligned to draw not just the shadow of the moon but the many shapes of Earth and life, of densely concentrated and shaped space and time.

Yet as humans explored the order of the cosmos they also discovered realms of chaos more powerful and unappeasable than any dragons trying to devour the sun. The mammalian progress recorded in these rocks had happened only because an asteroid had crashed into

Earth and inflicted massive chaos—tsunamis, fires, an atmospheric eclipse, climate change, ecological collapse, and mass extinctions, including the dinosaurs who had ruled Earth for more than a hundred million years. The asteroid had cleared the stage for mammals to thrive, yet they would always be indebted to chaos: they would walk with the calcium of dinosaurs, stare at the world through Chicxulub craters, and feed their young with a milky way that hid black holes. In asteroid dust they had practiced the footprints they would one day make upon moondust. And this asteroid impact was but a bee sting compared with the collision that had created the moon and remade Earth, cosmic violence that would continue and leave the moon saturated with craters and debris. This violence had also broken out on Earth as volcanic eruptions, floods, and mudflows, all of which were recorded in this rock with the animals they killed. The chaos that made moon craters and Earth volcanoes also made wounds streaming with blood. It planted within every creature a thermodynamic black hole that demanded to be fed, demanded a daily lottery of searching, demanded that animals murder one another, demanded that animals experience the universe through veils of fear and pain. The evolution recorded here was partly the progress of weaponry, of better teeth and claws for killing, better eyes for surveillance, better limbs for chasing or fleeing. Now this arms race had filled Verdun with craters and arrived at jets and missiles and nuclear warheads. The astronauts had earned their flights to the moon by being the masters of the newest weaponry, and they rode rockets designed for war. Some of the astronauts had flown bombers loaded with nuclear bombs, not just asteroids but supernovas ready to fill Earth with craters far larger than Verdun craters, ready to turn Earth into the moon.

The astronauts had brought with them to the moon

their long, heavy, dragging Earth tails, their biological natures and human identities, which cocooned them like spacesuits against being fully impacted by the moon. Earth had separated itself into bodies that regarded separation as their truest identity, that defined themselves by what made them different from the worlds around them. Separate species did not identify with one another, nor separate tribes, nor separate towns or families or bodies. Intensely social animals, humans defined themselves first of all by their roles within human society. Thus the astronauts did not perceive themselves as essentially "life" or "Earth" or "cosmos" or "Earth become alive." They did not perceive the moon landing as "Earth meeting moon" or "living world meeting dead world" or "Earth meeting its past." They perceived it first through their individual identities, their separate names and experiences and careers and ranks, feeling rivalries and pride even here on a world totally indifferent to human identities. They perceived it as a tribal event, one tribe's assertion of superiority over other tribes. They perceived it as a technological event, a proof of human cleverness. If they were feeling especially generous, they perceived it as a human event, an expansion of the realm and abilities of the human race—the one small step and the one giant leap remained the striving of an animal for a higher fruit.

Yet whatever their motivations, humans had taken a giant leap out of the identity boxes that had long confined them, a leap into far larger circles. The lunar lander was the golden key that fit into the lock of the cosmos and opened its door wide. The first machine ever designed to fit humans to another world, the lander sitting on the moon was physical proof of what had long been a mere idea, often a remote idea, that humans were part of a vast cosmos. To reach the moon humans had needed to obey the cycles that turned the solar system and brought full moons and eclipses. A human standing on the moon was a violation of normal human identities and an expansion of human contexts. Humans may have wanted to make the moon a part of their world, but instead the moon was inducting humans into its far larger and more powerful world. The planet Earth floating as a little ball in the moon's sky redefined Earth and humans as much smaller than they had imagined, invisible to the cosmos. Earth was not the endless landscapes humans had always experienced. Earth was one little ball among many. Yet Earth was the genius planet, the incredibly generous planet, one ball of life among many dead ones. This reality was so powerful that even the steely and inarticulate astronauts could not help feeling it, sometimes deeply, and most tried to give it words, and some did well at it.

The images of Earth in space and of humans on the moon could readily lead to a further refocusing of human context and identity, not a shrinking but an expansion, for when humans overcame their sense of separation from the cosmos they might see that they belonged to it not just as a physical object but as a creation and an embodiment. They would see their cosmic identity, that in the society of stars and planets and empty space they were the cerebrum of the convoluted nebulae, the celebrant of the universe's long journey out of chaos. And beyond that, beyond all roles and becomings, humans might glimpse in the floating, unsupported, ghostly, impossible Earth the glow and strangeness of existence itself.

I climbed a tall ridge behind the park visitor center, combing against grasses that attached their seeds to my socks and shoelaces, their hopes of surviving long into the future. By climbing high, I was hoping to get a better view of the moon's shadow moving across the landscape before me, the valley of the John Day River and its opposite ridges, including a peak hundreds of feet high called Sheep Rock, which had reminded local sheep ranchers of a resting sheep, though I wasn't so sure, so sheepish. I mainly saw many bands of color, odd greens and blues and pinks, minerals forged deep within volcanoes and sprayed across the land again and again, many pages that made this one of the world's best books for reading the details of mammal evolution. The bands of color were capped by bands of black basalt lava flows. All the bands had been tilted northward, tiled by the same tectonic forces that had created these volcanoes, and then they had been eroded into a valley by the river, whose riverbed sands and gravels also held odd colors. This surreal landscape seemed a good screen on which to watch such a surrealistic event, which would bring its own odd colors. I spread a blanket on the slope and stretched out to wait. The grasses would be happy that I would pluck out their seeds two dozen miles away, in my mountaintop campsite, that evening.

I had arrived at six in the morning, three hours before the eclipse, so I watched the river flowing and brightening, and Sheep Rock changing colors and shadows. I watched birds flying and grasses swaying and humans arriving. I also tried to watch the celestial motion of the land. I could know the exact beginning time of the eclipse, down to the second, and its exact duration not just here but at every spot on the continent, and its exact degree of darkness at various latitudes, because of the reliable motions of the solar system. The moon might be a crude and oblivious mass, 81 quintillion tons of it, powerless to control its own motions, but it belonged to a covenant of roundness, worshipping roundness with its shape, spin, and orbit, with precise rituals of motion. I imagined the first touch of the moon's shadow heading steadily this way, over the ocean, hitting the beach, encouraging the fog, following the rivers, climbing the mountains and volcanoes, crossing into the drylands where I was waiting. All the other people who had gathered here, on time, also seemed entirely confident of the solar system's order, that the moon's shadow would arrive exactly on time. And here it was: the first dark nibble out of the sun. At first it was so subtle, I wondered if I was imagining it. The nibble grew slowly, becoming a black curve upon the light, growing steadily for more than an hour.

This time I wanted to watch the land changing its light and shadows and colors as the eclipse proceeded, and I had a great setting for that, though I doubted it was the right kind of promontory from which someone could see the moon's shadow rushing across the land like a tsunami. For awhile, there was no obvious change on the ground. The river offered the first evidence: its sparkles were fading. The ground and air gradually turned into a strange twilight. The moon was walking upon Earth, heading this way.

In this eclipse I could see something that previous generations had not seen. I could see the moon holding six gold and silver medallions that Earth had pinned upon it, medallions surrounded by footprints, footprints already fossilized as thoroughly as the fossils beneath me, footprints by which the moon had been inducted into the circle of life, even if these were tiny footprints compared with the moon's massive stride through space.

I did feel small against the eclipse and the vast bodies and cycles it embodied and made more palpably real and powerful. It was not right that it took an eclipse to make the cosmos feel real, for of course the cycles of the cosmos were visible all the time, with every "sunrise" and "sunset" and phasing of the moon. But the human mind, with its deep acclimation trackways and zombie sense of wonder, refuses to notice anything odd or wonderful about the reliable cosmos, so it takes a violation of the natural order to startle us. I was feeling an awe that seemed to go deeper than normal human emotions, that might be flowing out of the mammalian skulls I had seen in the museum, not so fossilized after all. It was a delicious awe, an accurate smallness, yet one that held animal surprise and fear. Perhaps in the museum the fossils were scurrying for shelter. At least, it seemed the birds around me were falling silent.

The moon continued advancing, the sun shrinking. Ninety percent, ninety-five. At the last moment I took off my eclipse glasses and looked at Sheep Rock and the river valley, hoping to see the shadow tide racing along, and at least I saw the ground darkening quickly and oddly.

Above, a few planets and stars were peeking out. They were, of course, always there, forgotten by humans who imagined that the universe consisted of daylight and human activities.

The countdown hit zero, and the moon's shadow was complete, engulfing the sun and the ground and me. I looked at the masked sun and saw its blackness and its always-hidden corona shining bright, streaming outward, full of flickerings. It was a revelation. I felt beauty, magic, unreality, and reality. The corona was a glorious halo, perhaps stirring from our mythic imagination symbols of perfection, like the angels we called upon to lighten our troubling existence. I tried to shut up my symbols and science and thoughts and simply see and feel and enjoy it.

Yet I could not forget that totality lasted only two minutes and a few seconds, and it was impossible not to feel time passing. It passed both very slowly and far too quickly. I wanted to hold on to the moment, even if holding on was a denial of the moon's message that humans belonged to great and endless motions.

Motion and gravity called, and the moon slipped off the sun, creating a gleaming diamond ring. I had to put my eclipse glasses back on to hide from my creator star.

As the moon moved onward, I watched the cosmos flowing, its massive bodies and forces and cycles, and I marveled that they had flowed into geological forms and forces, becoming tectonic plates and volcanoes and ash strata and rivers, and flowed beyond that into all the shapes of life. I felt the cosmos within me. I marveled that craters could become eyes.

The birds began singing again. The great shadow was melting into thousands of Earth shadows. The light and colors began returning to normal. Yet I hoped this eclipse would disallow the moon and sun and Earth from returning to their old normality. The moon had sliced through human normality with a laser of darkness, with the reality of the cosmos. It had demanded that humans stop and pay attention. It held up a stone mirror in which we could see ourselves. With its eerie lights and colors and darkness, it had revealed our own true strangeness.

Don Lago is an award-winning writer who has published more than fifty nature and astronomy essays in national magazines and literary journals. He has authored several books, including most recently *Grand Canyon: A History of a Natural Wonder and National Park.* He lives in Flagstaff, Arizona.

Printed and bound by PG in the USA

USA2019PGIL